Africa Now

Africa Now is an exciting new series, published by Zed Books in association with the internationally respected Nordic Africa Institute. Featuring high-quality, cutting-edge research from leading academics, the series addresses the big issues confronting Africa today. Accessible but in-depth, and wide-ranging in its scope, Africa Now engages with the critical political, economic, sociological and development debates affecting the continent, shedding new light on pressing concerns.

Nordic Africa Institute

The Nordic Africa Institute (Nordiska Afrikainstitutet) is a centre for research, documentation and information on modern Africa. Based in Uppsala, Sweden, the Institute is dedicated to providing timely, critical and alternative research and analysis of Africa and to cooperating with African researchers. As a hub and a meeting place for a growing field of research and analysis, the Institute strives to put knowledge of African issues within reach for scholars, policy-makers, politicians, the media, students and the general public. The Institute is financed jointly by the Nordic countries (Denmark, Finland, Iceland, Norway and Sweden).

www.nai.uu.se

Forthcoming titles

Fantu Cheru and Renu Modi (eds), *Agricultural Development and Food Security in Africa: The Impact of Chinese, Indian and Brazilian Investments*

Margaret C. Lee (ed.), *World Markets and Trading Regimes in Africa*

Amanda Hammar (ed.), *Displacement Economies in Africa: Paradoxes of Crisis and Creativity*

Titles already published

Fantu Cheru and Cyril Obi (eds), *The Rise of China and India in Africa: Challenges, Opportunities and Critical Interventions*

Ilda Lindell (ed.), *Africa's Informal Workers: Collective Agency, Alliances and Transnational Organizing in Urban Africa*

Iman Hashim and Dorte Thorsen, *Child Migration in Africa*

Prosper B. Matondi, *Zimbabwe's Fast Track Land Reform*

Prosper B. Matondi, Kjell Havnevik and Atakilte Beyene (eds), *Biofuels, Land Grabbing and Food Security in Africa*

Cyril Obi and Siri Aas Rustad (eds), *Oil and Insurgency in the Niger Delta: Managing the Complex Politics of Petro-violence*

Mats Utas (ed.), *African Conflict and Informal Power: Big Men and Networks*

About the authors

Maria Eriksson Baaz is associate professor at the School of Global Studies, University of Gothenburg, and a senior researcher at the Nordic Africa Institute, Uppsala, Sweden. Her research interests are African politics, security and development, post-colonial theory and gender. Recently she has focused on masculinity, militarization and defence reform interventions, with a particular focus on the Democratic Republic of the Congo. She is the author of *The Paternalism of Partnership: A Postcolonial Reading of Identity in Development Aid* (2005). She has also contributed to several edited volumes, such as the *International Handbook on African Security* (2012), and has written numerous policy reports. Additionally, her articles have appeared in leading journals, including *International Studies Quarterly*, *African Affairs*, *Journal of International Relations and Development*, *Journal of Modern African Studies* and *African Security*.

Maria Stern is professor of peace and development studies at the School of Global Studies, University of Gothenburg. Her research interests are security studies, the security–development nexus, politics of identity, and feminist theory. Recently she has focused on masculinity, militarization and defence reform interventions, with a particular focus on the Democratic Republic of the Congo. Maria co-edited a special issue on the 'Security–development nexus revisited' in *Security Dialogue* (2010). She is also co-editor of *Feminist Methodologies for International Relations* (2006) and the author of *Naming Security – Constructing Identity* (2005). She has contributed to several edited volumes, such as the *International Handbook on African Security* (2012), and has written numerous policy reports. Additionally, her articles have appeared in leading journals, including *African Affairs*, *Alternatives*, *International Journal of Peace Studies*, *International Political Sociology*, *International Studies Quarterly*, *Journal of International Relations and Development*, *Journal of Modern African Studies*, *Review of International Studies* and *Security Dialogue*.

Sexual violence as a weapon of war?

Perceptions, prescriptions, problems in the Congo and beyond

Maria Eriksson Baaz and Maria Stern

Nordiska Afrikainstitutet
The Nordic Africa Institute

Zed Books
LONDON | NEW YORK

Sexual violence as a weapon of war? Perceptions, prescriptions, problems in the Congo and beyond was first published in association with the Nordic Africa Institute, PO Box 1703, SE-751 47 Uppsala, Sweden in 2013 by Zed Books Ltd, 7 Cynthia Street, London N1 9JF, UK and Room 400, 175 Fifth Avenue, New York, NY 10010, USA

www.zedbooks.co.uk
www.nai.uu.se

Editorial copyright © Maria Eriksson Baaz and Maria Stern 2013

The rights of Maria Eriksson Baaz and Maria Stern to be identified as the editors of this work have been asserted by them in accordance with the Copyright, Designs and Patents Act, 1988

Set in OurType Arnhem, Monotype Gill Sans Heavy by
Ewan Smith, London
Index: ed.emery@thefreeuniversity.net
Cover design: www.roguefour.co.uk

A catalogue record for this book is available from the British Library
Library of Congress Cataloging in Publication Data available

ISBN 978 1 78032 164 6 hb
ISBN 978 1 78032 163 9 pb

Contents

Abbreviations and acronyms

COIN	counter-insurgency
DRC	Democratic Republic of the Congo
FAC	Forces Armées Congolaises (Armed Forces of Congo)
FAPC	Forces Armées du Peuple Congolais (People's Armed Forces of Congo)
FARDC	Forces Armées de la République Démocratique du Congo (Armed Forces of the Democratic Republic of the Congo)
FMLN	Farabundo Marti National Liberation Front (El Salvador)
ICTR	International Criminal Tribunal for Rwanda
ICTY	International Criminal Tribunal for the former Yugoslavia
INGO	international non-governmental organization
IR	international relations
MLC	Mouvement pour la Libération du Congo (Movement for the Liberation of Congo)
MSF	Médecins sans Frontières
NGO	non-governmental organization
OCHA	(UN) Office for the Coordination of Humanitarian Affairs
PRIO	Peace Research Institute Oslo
RCD	Rassemblement Congolais pour la Démocratie (Congolese Association for Democracy)
Sida	Swedish International Development Cooperation Agency
SSR	security sector reform

Acknowledgements

The bulk of the research for this book was made possible through several generous grants from the Swedish International Development Cooperation Agency (Sida). We also received a grant for research assistance from the Gothenburg Centre of Globalization and Development, University of Gothenburg. Chapter 2 of this book emerged from research conducted under the auspices of the project Arms Against a Sea of Troubles at the Peace Research Institute Oslo (PRIO). We are grateful for this support.

Additionally, we owe many thanks and much appreciation to people within the Congolese security forces, who greatly facilitated our research. We are especially indebted to the soldiers and officers of the Congolese armed forces, who generously shared their experiences and thoughts with us. We also thank members of various organizations working in the DRC, who spoke with us about their experiences and points of view.

Hanna Leonardsson, who acted as research assistant throughout the writing of this book, made the seemingly impossible possible, through her patience, hard work and resourcefulness. Hanna Leonardsson, Maria Malmstöm and Molly MacGregor assisted in collecting various forms of data on the discourse of 'rape as a weapon of war' in policy texts and media. Many thanks!

We also want to extend our gratitude to those who have provided pertinent, insightful and valuable comments on the manuscript: Paul Higate, in particular, earns special thanks. We also thank Mikela Lundahl, Stina Hansson, Véronique Pin-Fat, Kaia Stern, Mats Utas, Judith Verweijen, Marysia Zalewski and members of the Global Gender Studies research group at the School of Global Studies, University of Gothenburg.

Last, but certainly not least, we would like to thank the editorial team at Zed Books. In particular, we thank Kika Sroka-Miller for her comments on the manuscript and Ken Barlow for his never-ending patience and belief that we would indeed eventually finish this book.

We dedicate this book to our children: Kiwa and Emmanuel Eriksson; Alexander and Andreas Stern; and bonus children: Oskar, Kåre, Astrid and Ylva Fridell – and to 'mormor' Ingela Eriksson and Erik Fridell, whose support makes our work possible.

Introduction

> It has to be understood that this is a security problem, not just men behaving like men. It's not an inevitable consequence of war – it's something that is planned. It can either be commanded, condemned or condoned. We need to say that we can stop it. It's not inevitable. (Margot Wallström, cited in Crossette 2010)

Finally, the international community has recognized conflict-related sexual violence as an important global security problem. Indeed, the notion that rape is a weapon of war that warrants global attention has become commonplace in media reporting and policy analysis. Despite the often horrific violences it documents, the prevailing and now familiar story of wartime rape is a story that fills us with hope. While we may be intermittently confronted with terrible images of rape survivors in ghastly conditions on our television screens or in the newspapers we read, we are nonetheless slightly comforted. After years of silence and neglect, the ills of rape in war are finally being named. Redress for victims of rape has become a high priority, and, we are reassured, the systematic and widespread scourge of sexual violence will someday be halted, or at least seriously hindered. Sexual violence as a weapon of war has at long last begun to receive the attention it warrants, given the suffering its victims endure and the societal harms it occasions. Indeed, we are confident that a crucial key to further understanding and eventually redressing conflict-related sexual violence has been obtained through its being recognized as an acute and serious global security problem, as a 'weapon of war'. Yet, in the midst of our horror over the atrocity of rape, the sense of feminist success that rape and its sufferers are rendered visible, and the relief that *something* is finally being, or about to be, *done*, we feel a growing unease. This unease is the subject of this book.

First, let us explain the success. While the history of rape in war is as long as the history of warring itself, until recently it has been largely ignored. Rape was generally treated as if it were an 'unfortunate by-product' of warring (Seifert 1994), warranting little if any attention in the 'high politics' of global and national security. However, after far too many centuries of silence and neglect, the pressing issue of sexual violence in war has now finally been recognized in the wake of the international recognition of the mass rapes during the armed conflicts in both Rwanda (1994) and Bosnia-Herzegovina

(1992–95). Much policy and media attention has since been paid to the scourge of conflict-related sexual violence, particularly the role of sexual violence in the conflict in the Democratic Republic of the Congo (DRC).

Hence, since 1993 there has been a marked shift in the ways in which sexual violence has been framed in the global policy debate. Dominant understandings have moved from perceiving rape in war (if remarked on at all) as a regrettable but inevitable aspect of warring, to seeing it as a strategy, weapon or tactic of war, which can be prevented. Indeed, several United Nations Security Council Resolutions[1] and the appointment of a Special Representative on Sexual Violence in Conflict have confirmed the United Nations' commitment to combating conflict-related sexual violence.

The notion that rape is a (systematic) weapon of war whose use can ultimately be hindered depends upon a narrative or a frame of understanding which assigns particular meanings to rape in war, as well as to rapists and the victims/survivors of rape. The story told and retold about rape and its subjects in the media and policy reports, as well as in much academic writing, makes good sense. Indeed, the compelling and seemingly cohesive narrative of rape as a (gendered) weapon of war is revolutionary in its global appeal and exemplary in its successful call for engagement to redress the harms of rape – especially in the case of the DRC.

Yet this triumph also elicits our concern. Simply put, our fear is that the dominant framework for understanding and addressing wartime rape has become so seemingly coherent, universalizing and established that seeing, hearing and thinking otherwise about wartime rape and its subjects (e.g. perpetrators, victims) is difficult. In other words, this dominant framework reproduces a limited register through which we can hear, feel and attend to the voices and suffering of both those who rape and those who are raped. Despite its progressive appeal, political purchase and success in bringing attention to many who suffer, the newly arrived accomplishment of recognizing rape as a weapon of war thus may also cause harm.

Ours is surely not a unique concern.[2] On the tails of accomplishments like the UN Resolutions noted above come also a host of problems and dilemmas. Any framework for understanding and redressing complex problems, such as sexual violence in war, is bound to be limited and limiting. That said, in order to move or peek beyond these limits, we need to explore them: how have they been constructed? What purposes do they serve? Indeed, it is the call to explore the limits of the prevailing ways of thinking about sexual violence in war which prompts us to write this book. Our critical inquiry, however, is not intended to be damning, but instead it is offered as a contribution to a healthy and considered reflection of the contemporary politics of framing sexual violence in war (Butler 2009). Hence, in this book, we critically engage with dominant understandings of, as well as policy solutions aimed at redressing, sexual violence in conflict and post-conflict settings. In short, the book explores the main story of Rape as a Weapon of War: its underlying assumptions, ontologies, composition and limits.

What interests us is the ways in which rape is imbued with meaning in the governing discourse about sexual violence in warfare through certain 'grids of intelligibility'.[3] These grids of intelligibility circumscribe what can be said about rape in war, as well as what kinds of subjects can exist in the main storyline of Rape as a Weapon of War. In the global frenzy to frame 'the disaster' of sexual violence in comprehensible terms, we argue, nuance and complexity are sacrificed and violences are both produced and reproduced (Dauphinée 2007; Zizek 2009).

In different ways in the following chapters, we therefore query the seemingly cohesive and certainly compelling narrative of wartime rape, unpack its prevailing logics, explore its limits, and examine its effects. In so doing, we address some of the dilemmas and thorny issues inherent in the success of the 'arrival' of sexual violence on the global security agenda. While the majority of the book (Chapters 1–3) is preoccupied with interrogating and unpacking the dominant narrative about wartime rape as a 'weapon of war' as articulated in academic, policy and media texts, the last chapter also explores some practical interventions that have emerged in light of this narrative. Hence, we not only query how the discourse of Rape as a Weapon of War is constructed through, among other things, the exclusion of potential stories and voices, we also interrogate the ethico-political implications of interventions aimed at combating this violence.

Our critical reading as a whole rests upon explorations in several interwoven, overlapping and related registers. We will return to a description of each chapter below. Here, we first outline the moves the book makes in broad strokes.

The following two chapters are explicitly about the storylines that fill the Rape as a Weapon of War discourse with meaning. We begin our journey by exploring the interconnections between sex, gender and violence as a way of querying the underlying logics, or narratives, upon which the Rape as a Weapon of War discourse rests. In particular, we explore two deeply intertwined, generalized narratives: the story of sexual violence in warring as rooted in nature and biological urges (the 'Sexed' Story, as we call it) and the 'Gendered' Story which has supplanted it in terms of appeal and purchase. As we shall see throughout the book, the 'Gendered' Story explicitly overlaps with and performs important functions in the story of Rape as a Weapon of

War, while the 'Sexed' Story informs the Rape as a Weapon of War discourse through its exclusions and racialized spectres. Indeed, the dominant framing of Rape as a Weapon of War cannot be understood outside the 'Gendered' Story (and, again, the excluded 'Sexed' Story). The 'Gendered' Story will show that it is the gendering of the perpetrators and victims of war which constructs rape as weapon via its power and efficiency. Moreover, the storyline of rape in war as gendered (rather than 'sexed') performs a crucial function in reversing the idea of rape as an unavoidable consequence of war. Importantly, we query the assumptions (or ontologies) that underpin this understanding of sexual violence as gendered (instead of sexed) and ask who and what is silenced or dehumanized?[4] What other voices whisper in the margins of the central attraction? What stories can we hear and not hear?

Another entry point into our interrogation of the dominant framings of wartime rape is through a more specific unpacking of the discourse of Rape as Weapon of War and the crucial notion of 'strategicness'[5] upon which this discourse rests. The strategic use of rape is often presented as somehow self-explanatory through its implied universalized storyline of gender and warring. What sorts of assumptions are needed to make this claim/explanation possible? And why is this framing of sexual violence so seductive and so prominent? What kinds of subjects does it produce and exclude?

As we argue throughout the book, the pervasive aspect of the Rape as a Weapon of War discourse rests, largely, on its promises of change and the policy implications it offers in writing rape in war as preventable; as an abhorrent condition that can be treated. After years of silence and portrayals of rape as unavoidable, this narrative promises a brighter future for sexually abused women (and men) in conflicts. The Rape as a Weapon of War discourse is decidedly policy friendly, lending itself to the necessary reductionism for arriving at viable policy goals, which can also be placed in a results-based framework. Hence, in the urgency to redress sexual violence within global security policy, a framework for understanding that is seemingly cohesive and universal emerges that – more often than not – poorly reflects the realities of the complex warscapes[6] in which it is applied. Furthermore, through its universalizing narrative, the discourse may conceal and exclude subjects and accounts that could improve understanding of or add additional knowledge about how and why sexual violence in warring occurs, as well as what it may mean to those who are subjected to it.

As is apparent from the preceding discussion, this book explores stories, or ways of framing rape, rather than offering explanations for why sexual violence constitutes a common act of violence in many conflict settings. However, while we unpack dominant understandings (rather than provide explanations for why rape takes place), we also invite the reader to consider some alternative understandings of sexual violence. By highlighting that which is excluded and silenced in the prevailing storyline – by revealing its lacunae and its limits – we draw attention to additional ways of understanding sexual violence that are relevant in warring contexts but have been excluded by the dominant discourse. Drawing upon insights collected from the sociology of violence and the military, as well as research conducted in the DRC (see below), we highlight frameworks for understanding violence, as well as aspects of military structures that are silenced in the dominant story of rape. In some contexts, such as the conflict in Bosnia, sexual violence in war seems to be best understood as a conscious strategy to fulfil political and military goals; in some military structures, orders are effectively enforced down the chain of command so that such a strategy is (more or less) effectively implemented. However, we discuss how sexual violence can *also* reflect the opposite: the breakdown of chains of command; indiscipline, rather than discipline; commanders' lack of control, rather than their power; the micro-dynamics of violent score-settling, rather than decisions of military and political leaders engaged in defeating the enemy.

As noted above, our exploration into the underlying logics and scaffolding of the Rape as a Weapon of War discourse emerges out of a concern with the ways in which a generalized story of rape in war limits our abilities to analyse and redress instances of sexual violence in specific warscapes, as well as to attend to the people whose lives are circumscribed by such violence. We therefore also contemplate the politics of humanitarian engagement. In particular, we consider the ethics and dilemmas of trying to combat sexual violence and to alleviate the plights of the victims of sexual violence and ask the following questions: What does the new-won attention to wartime sexual violence fail to deliver to women (and men) in post-conflict settings (in this case the DRC)? What relations of power are concealed in the politics of solidarity and humanitarian work? And finally, what are the politics of applying such a critique in such a highly charged setting, where lives are highly vulnerable and precarious?

Learning from the DRC: the so-called 'rape capital of the world'[7]

The Democratic Republic of the Congo (DRC), long known by many as 'the heart of darkness' (Conrad 1990 [1902]), has been redubbed the 'rape capital of the world'.[8] Indeed, the DRC has become infamous globally through reports on the alarmingly vast amount of sexual violence that has accompanied devastating armed conflicts. While other forms of violence have also been committed on a massive scale, it is sexual violence which has attracted the lion's share of attention, especially among 'outside' observers. This *singular focus* on sexual violence has been reflected in the number of reports, articles, news clips, appeals and documentaries dealing specifically with the issue of rape. Other forms of violence – mass killings, systematic torture, forced recruitment, forced labour and property violations, etc. – are committed on a massive scale but receive far less attention and resources.[9] Sexual violence has been described as the 'monstrosity of the century' (Li Reviews 2008), 'femicide', a 'systematic pattern of destruction toward the female species' (Eve Ensler, cited in Kort 2007), 'incomprehensible' (Nzwili 2009), the 'worst in the world' (Gettleman 2007), etc. Numerous journalists, activists and representatives of diverse international organizations and governments have made pilgrimages to the DRC to meet and listen to survivors first hand. Arguably, with this attention, 'rape tourism' has been added to what has come to be known as 'war zone tourism' (Eriksson Baaz and Stern 2010).

While this book explores broad questions, fears and concerns about the framing of sexual violence in warring more generally, it is grounded in extensive first-hand research in the DRC warscape. Throughout the book, we therefore draw upon the site of the DRC as examples of, or points from which to pose questions about, the more general renditions of wartime rape. We want to emphasize, however, that our intent here is not to offer a comprehensive understanding of wartime rape in the DRC. Our analysis draws upon – and problematizes – our knowledge of the DRC warscape, but goes beyond the DRC as a case. It is therefore relevant for understanding the framing of sexual violence in conflict and post-conflict settings more generally. Furthermore, the considerable attention paid to sexual violence in the DRC, which is reflected in the interventions of various international actors, renders the DRC a particularly good case from which to learn. Our knowledge of the workings of the armed forces and the problematics of sexual violence in the DRC therefore provides a fruitful point of departure from which questions can be posed both in general terms and in relation to other specific conflict settings.

The references to the DRC that appear throughout this book emerge from several interrelated research projects that we have conducted. In particular, we draw from a research project exploring gender in the military, which is based on interviews with soldiers and officers in the Congolese national armed forces (FARDC).[10] The interviews addressed how the soldiers themselves saw their role in the armed forces, as well as in relation to civil–military relations. We asked them about their understandings of what it meant to be a 'good soldier', and of masculinity and femininity in relation to soldiering. In particular, we focused on the reasons that soldiers gave for why rape occurs and on what they told us rape is or means. We did so in order to query some of the governing discourses, and the subject positions designated through the workings of these discourses (e.g. what it means to be a 'soldier' or a 'man' within the FARDC), which were reflected, reproduced and renegotiated in their narratives. Indeed, our extensive experiences researching wartime sexual violence in the DRC, and importantly the questions we have subsequently posed concerning our own research process and results, are the impetus behind the writing of this book. Let us explain further.

By attending to the voices of the soldiers who speak about perpetrating rape, we had hoped to find a venue other than that commonly traversed for understanding the occurrence of (sexual) violence in the DRC. Yet, when we attempted to complicate and disrupt the main storyline of rape that we had been conditioned to hear and to tell, we were thwarted by its strong hold. The grids of intelligibility available to us as practised scholars, well versed in IR feminist theory and participants in public political debate, left us bereft of a lexicon for properly hearing and writing about rape differently – in a way that did justice to the stories the soldiers told us. Indeed, as scholars thinking, writing and teaching on gender and war, we have participated in reproducing these storylines (see Stern and Zalewski 2009). Surely, our intended story of rape was precluded by the assumptions about ethics, subjectivity and violence that framed our question of 'why soldiers rape?' in the first place. We continued nonetheless to bang our heads against the limits of possible imaginings, and were frustrated in our inevitable failings and complicity in violent reproductions of rape, rapists and victimhood.

We also draw upon a smaller research project entitled 'Gender-based violence: understanding change and the transformation of gendered discourses'.[11] This project was based on interviews with national and local organizations in the DRC, working in the area of women's rights, with the aim of examining how their understanding of sexual violence and gender relates to that of international actors in the field. Again, in making sense of women's and NGOs' stories about their fears, needs and survival strategies, we sometimes found ourselves adrift without a comfortable language for listening to or writing about their concerns.

Some additional notes on theory and methodology

Theoretically and methodologically, this book is a bit unruly. In addition to drawing on diverse research areas, it also draws on scholarship that rarely meets but instead tends largely to ignore each other's writings.[12] While the book can be situated in feminist theory, it reads both *with* and *against* feminist analyses of the interconnections between gender, warring, violence and militarization. One aspect of 'reading against' is that we draw upon literature that seldom features in feminist texts: military sociology. Through a seeming 'guilt by association' logic (where citing military sociology implies that one is associated with militaristic goals), military sociology has been largely ignored in much feminist research.[13] While there certainly are some valid grounds for this exclusion, we believe that this body of research can provide important insights that are otherwise neglected in the dominant story of wartime rape. Particularly, much work within military sociology highlights and seeks to arrive at remedies for the failures of military institutions, often aiming at increasing their efficiency. Consequently, and in contrast to the dominant story of wartime rape, this literature tends to establish and explore the incompleteness of military structures. Often such literature, as we shall see, points to the failings of military organizations to work according to the ideals of discipline, hierarchy and control. By neglecting this literature and by not acknowledging these 'failures' (but instead portraying the military institution as the rational war machine it aspires to be), the Rape as a Weapon of War discourse, in a twist of irony, tends to mimic the adulating self-image cultivated by its rejected militaristic Other.

Moreover, the book also draws upon post-colonial theory. While post-colonial theory offers vital insights into the general story of rape in war, it is (unfortunately) indispensable in grasping the framing of sexual violence in the so-called 'rape capital of the world'.

The book is eclectic also in terms of methods. Chapters 1 and 2 are based on discourse analysis (i.e. focus on the construction of meaning), although Chapter 1 is written in a much looser exploratory analytical form than Chapter 2, which follows a stricter form of discourse analysis. In Chapters 3 and 4, we offer a literature overview and analysis; in addition we present data on events, processes and consequences of interventions in the DRC warscape, as well as the workings of military structures.

Before we offer a brief synopsis of each chapter, let us pause to clarify what we mean when speaking of the Rape as a Weapon of War narrative as a discourse. Analysing the dominant narrative of wartime rape through the tools of discourse analysis helps us to unpack and make sense of the ways in which the storyline has reproduced knowledge about rape, as well as its subjects (e.g. perpetrators and victims, as well as policy practitioners and researchers/experts). We understand discourses to be historically, socially and institutionally specific structures of representations, and partial, temporary closures of meaning (see Eriksson Baaz 2005). Importantly, discourses function by giving a semblance of cohesion, order and closure. They make *sense*.[14]

Discursive structures can be understood as a system of differences in which the identity/meaning of the elements is purely relational.[15] Understood in this way, a discourse does not contain a given stable definitive content, but requires that which it excludes (and which threatens its hegemony over meaning/identity) as integral to its structure in order for it to make sense.[16] Further, 'any seemingly coherent representation is always an unstable configuration insofar as "it" is constituted by, and indeed haunted by, that which is excluded. These hauntings, or constitutive outsides, are forever present' (Pin-Fat and Stern 2005: 29; Pin-Fat 2000). This is what we mean when we refer to the 'hauntings' of excluded stories or subjects throughout the book. Furthermore, there are many competing discourses at play in any discursive field; within any discourse, traces of other competing discourses persist. Consequently, discourses (even dominant ones) are merely temporary fixations, which, by necessity, are never complete, although they often masquerade as a universal totality. Instead, discourses are always inherently unstable, because of their relation to other discourses and their being constituted through difference and exclusion. Discourses therefore demand continual reinforcement because of the inevitable contestations they incite (Weldes et al. 1999: 9). They therefore can never fully succeed in hegemonizing meaning. Therein lies the continual possibility for contestation of dominant discourses and the ideologies or logics that underwrite them – a possibility which we embrace and explore in the different chapters of this book. Hence, using our methodological toolbox of discourse analysis, we are thus able to better glimpse how meaning is being produced in the discourse of Rape as a Weapon of War and the 'Gendered' Story of rape upon which this discourse rests.

Outline of the book

In Chapter 1, 'Sex/gender violence', we depart from our experiences of researching rape in the DRC and argue that the dominant and seemingly progressive frame of seeing, listening to and understanding wartime rape, when probed, reveals a host of unexamined effects. We set the stage for the subsequent analysis (particularly in both the remainder of Chapter 1 and Chapter 2) by offering a reading of the dominant narratives that frame possible understandings of sexual violence: first the 'Sexed' Story of wartime rape, followed by the 'Gendered' Story, which has seemingly replaced it. The chapter then explores how the 'Gendered' Story (and the 'Sexed' Story that haunts it) produces sexual violence as both normal and 'abnormal', and fundamentally different from and outside of other forms of violence, which are presumed to be ungendered. Both of these moves (rendering sexual violence normal and abnormal simultaneously), we argue, ultimately contribute to dehumanizing those who rape and also ultimately those who are raped. It is therewith difficult to see and hear those who are subject to sexual violence in ways that we do not expect. We therefore briefly explore some of the uncomfortable subjects, who/which do not neatly fit into the dominant framing. In light of these 'uncomfortable subjects' we reflect on the ethico-political implications of writing about those who rape in the DRC, instead of about their victims. We explore the conundrum of complicity in researching violence and those who commit violence and explore the thorny questions of the ethics, dilemmas and fears that arise when attempting to understand how rape becomes possible from the perspective of those who commit these acts.

Chapter 2, 'Rape as a weapon of war?', offers a critical reading of the Rape as a Weapon of War discourse in order to make it visible and study its scaffolding (against the backdrop of our analysis in Chapter 1). In so doing, we identify four nodal points[17] that are central to producing meaning and coherence: strategicness, gender, guilt/culpability and avoidability. What sorts of assumptions are needed to make the claim that rape is a weapon or strategy of war? And why is this framing of sexual violence so seductive and so prominent? We ask these questions in order to better understand its appeal in the face of the violence of widespread and brutal conflict-related rape. This appeal, we suggest, resides in its inchoate promise that: the bestial violent sex evoked in the 'Sexed' Story and (ironically) reproduced in the 'Gendered' Story can be hampered; criminals will come to justice; wartime rape can be eradicated, or at least largely prevented or avoided; and sexual violence can be controlled, managed and depoliticized.

Chapter 3, 'The messiness and uncertainty of warring', is of a slightly different character to the preceding ones. Here we attend more specifically to the nodal point of strategicness in the story of Rape as a Weapon of War. Drawing upon insights collected from the sociology of violence and the military, as well as our (and others') research in the DRC, we explore the notion of rape as inherently strategic in warring. The aim of this chapter is to highlight some aspects of military organizations and warring that tend to be rendered invisible in the story of the strategicness of rape. We address three aspects in particular. First, we attend to the discursive nature of strategy and demonstrate the ways in which notions of military strategicness, including the strategicness of sexual violence, vary depending on military contexts. Secondly, we turn to the workings of military institutions and highlight the fact that military institutions rarely embody their ideals of discipline, hierarchy and control. Rather than reflecting strategic action, sexual violence in war can also reflect the fragility of military structures and hierarchies. Thirdly, we discuss how the 'messy' realities of warring trouble notions of rape in war as a strategic weapon of war by attending to the micro-dynamics of warring.

In Chapter 4, 'Post-coloniality, victimcy and humanitarian engagement:

being a good global feminist?', we shift our focus on to the politics and ethics of (international, external) engagement for redressing the harms of wartime sexual violence. We do so by providing a post-colonial reading of the global battle to alleviate the suffering of the raped women in the DRC. Specifically, we argue that the massive engagement in the plight of Congolese rape survivors offers an illuminating example of the re-enacting of the white wo/man's burden to 'sav[e] brown women from brown men' (Spivak 1988: 297).[18] In this chapter we also discuss some of the unintended consequences of the interventions designed to combat the so-called 'rape epidemic' and attend to its victims. We explore how a singular focus on sexual violence within a very wide repertoire of human rights abuses occasions selective listening and blinded seeing, as well as, more concretely, a 'commercialization of rape'. However, as the interventions themselves are problematic, so also is the critique of these interventions; in whose interest is this critique really articulated? What are the potential consequences/possibilities/risks of such critical interventions? How is the dominant story of wartime rape manifested in practical interventions aimed at redressing sexual violence? And with what consequences? In sum, we find that there is indeed ample cause for hope beyond the Rape as a Weapon of War discourse.

In Chapter 5, 'Concluding thoughts and unanswered questions', we re-cap our main points of analysis and further reflect on the ethico-politics of research and humanitarian engagement on rape in armed conflict settings. Importantly, we also address our own complicity in relation to the discourses and practices that we have queried (and criticized) in this book and discuss the pitfalls and possibilities of critique. In short, we 'attempt[s] to look around the corner, to see ourselves as others would see us' (Spivak 1999: xii–xiii).

1 | Sex/gender violence

Introduction

[...] A lone woman [in the DRC] [...] was stopped by a government soldier. 'Where are you going?' he asked, smiling while cradling an automatic rifle in his arms. She told him the name of her village [...]. 'I want to sleep with you' was his only reply. Shoving her to the ground and pulling down his green fatigues, the soldier thrust himself between her resistant legs with low, forceful grunts. She closed her eyes, wanting to be anywhere else but there alone with him on the side of an isolated road. His sweat and body odor were suffocating her [...] (Horwood et al. 2007)

They came out of the forest. Men with guns appearing barely human to the frail, ageing woman who months later recounted her ordeal, bent double after surgery to save her womb. 'They didn't look like men. Their skin was covered in cuts. Their clothes were completely torn. They became someone else, not humans.' (McGreal 2008)

MALE CORPORAL A: We soldiers commit rape, why do we commit rapes? Poverty/suffering [*pasi*]. When we are not paid, or not paid at all. We are hungry. And I have a gun. In my house my wife does not love me anymore [*mwasi alingaka ngai lisusu te*]. I also have a wish to have a good life like you [*nakoma bien lokola yo*].

MARIA EB: But that is a different thing, no? I asked about rape, not stealing [*vol/viol*].

MALE CORPORAL A: I understand, I understand. I am getting to it. I am not finished yet. Rape, what is that? It is connected to all that – stealing, killing, it is all in that [*ezali nionso na cadre wana*].

MARIA EB: So, it is anger [*kanda*] then or what?

MALE CORPORAL A: Yes, it is anger [*kanda*], it is creating, the suffering [*pasi*] is creating ... You feel you have to do something bad, you mix it all up: sabotage, women, stealing, rip the clothes off, killing.

MALE CORPORAL B: You have sex and then you kill her, if the anger is too strong [*soki kanda eleki, obomi ye*].

MALE CORPORAL A: It is suffering [*pasi*] which makes us rape. Suffering. If I wake up in the morning and I am fine, I have something to eat, my wife loves me [*mwasi alingaka ngaî*], will I then do things like that? No. But now, today we are hungry, yesterday I was hungry, tomorrow I will be hungry.

They, the leaders/superiors [*bamikonzi*], are cheating us. We don't have anything. (Cited in Eriksson Baaz and Stern 2009)

I propose to consider a dimension of political life that has to do with our exposure to violence and our complicity in it, with our vulnerability to loss and the task of mourning that follows, and with finding a basis of community in these conditions [...] I propose to start, and to end, with the question of the human (as if there were any other way to start or end!). We start here not because there is a human condition which is universally shared – this is surely not yet the case. The question that preoccupies me in light of the recent global violence is, Who counts as human? Whose lives count as lives? And finally, what makes for a grievable life? (Butler 2004a)

We start this chapter with these accounts because we are troubled by the question raised by Judith Butler: what 'makes for a grievable life'? (See also Butler 2009.) Or, as Brassett and Bulley (2007: 3), similarly ask: What counts as 'meaningful suffering'? In reading the first two quotations, intrinsically we feel the suffering of the 'lone woman' and the 'frail, ageing woman'. They are the victims; they are the ones we care for. They are the human beings in the story that we weave – a story of sexual violence committed by 'beasts'. They are the ones that we (the international community?) should protect. They compel immediate action to shield them from the palpable danger of what Žižek calls 'subjective violence': we must *act*, 'do something' (Žižek 2009: 5)! Yet, in the next quotation, the soldiers' story invites us to move from their plight and to wonder: Is there also another story to be told? Can we also feel the suffering of he who rapes? Can we feel empathy, or the desire and capacity for engaging in 'empathetic cooperation' with him (Sylvester 1994)? And what might caring for him do to him, to the 'lone woman' or the 'frail, ageing woman' ... and to us?

As noted in the Introduction, this book emerges from the experiences from a research project on gender in the military in the DRC (Congo), and is based on interviews with government soldiers and officers. In our research, we focused on the reasons the soldiers give for why rape occurs and what they tell us rape is, as well as how they make sense of themselves. Yet, when talking to the soldiers we interviewed, rereading their recorded testimonies, analysing their texts, writing of what we learned, and speaking about our research to both the academic and the policy communities, we continually found ourselves grappling with the questions of how we write/speak of the human who commits acts of sexual violence when the available discourses for recognizing rape ultimately refuse his/her humanity. How do we recognize and relate to the face and the voice of those who commit rape so that we can differently ask how rape and rapists become possible?[1] And, how do we handle the dilemmas that doing so pose for us?

In light of these questions, this chapter attends to thorny questions of the ethics, dilemmas and fears that arise when attempting to understand how rape becomes possible, even necessary, from the perspective of those who commit these acts. In so doing, it also calls into question who 'we' (critical international relations scholars, feminists, those who act and feel on behalf of humanity – Edkins 2005a; Jabri 2007a) are in relation to the perpetrators of sexual violence. It prompts us to interrogate how the 'games of truth' with which 'we engage as we write of war ... form a complicity as well as the other of war ...' (Jabri 2007b: 70–1). Ultimately, by bringing these questions to the fore, it invites further reconsideration about sex/gender, violence, subjectivity and ethics in 'world politics' (Hutchings 2007; see also Schott 2003). The reconsideration we offer here sets the stage for our subsequent interrogation of the dominant framing of sexual violence as a weapon of war throughout the rest of the book.

Our main intent here is not to understand wartime rape in the DRC as such, but to use the site of the DRC, as well as more general renditions of wartime rape, to explore the broader questions, fears and concerns about the framing of sexual violence in warring more generally. Instead of taking efforts to understand (and see/hear) those who have been raped as our point of departure (as is commonly the case in accounts that aim to rectify the silencing of rape and its victims), we reverse the usual order. Specifically, we revisit the framing of sexual violence in warring from the vantage point of the unease and discomfort that its contours imposed on our hearing the 'rapists'' story.[2] (Later, in Chapter 4, we explore the unease and discomfort occasioned when we hold ourselves accountable to those who have been raped.) Indeed, perhaps because of our very unease and failures, we are compelled to explore the ways in which such framings circumscribe the ways we can see, hear and attend to the subjects wartime rape produces.

Hence, this chapter examines the dominant story of sexual and gender-based violence (the 'Gendered' Story, as we call it) by first exploring the main account that precedes and haunts it (the 'Sexed' Story), then querying its main plot, and questioning what it does to us and to its main characters. What subjects does it produce?; who is silenced?; what stories can we hear? The dominant and seemingly progressive frames of seeing, listening to and understanding wartime rape, when probed, reveal a host of unexamined effects. In the frenzy to 'frame the disaster' of sexual violence in 'comprehensible terms' (Dauphinée 2007: 86), we argue, we reproduce familiar discourses (with their inherent and even violent inclusions and exclusions), delimit the registers through which we can hear and respond to suffering, and probably cause harm (Žižek 2009).

The chapter proceeds as follows. In the remainder of this introduction we further explain why we focus on unease as a methodological inroad to our analysis. In the next section we offer a brief and generalized rendition of, first, the 'Biological Urge/Substitution Theory' (the 'Sexed' Story), and then the narrative explaining the interconnections between gender and militarization (the 'Gendered' Story), which is seen to supplant it, and which underwrites the dominant framing of conflict-related sexual violence. Here we offer some reflections on the supposed move from sex to gender. Next we query how wartime sexual violence is presented as a particularly heinous crime (as *sui generis*, of itself), casting rape and those who enact rape as exceptional, and simultaneously both normal and abnormal – yet in slightly different registers.[3] In the following section, we explore the uncomfortable subjects who prompt us to avert our gaze and cover our ears. We conclude by returning to the questions evoked by Butler's words and the unease, dilemmas and fears elicited by our attempts to write rape, rapists and victims otherwise (Butler 2004b: 93).[4]

Unease Let us step back a moment. Why a focus on our unease? Our unease has to do both with the discomfort caused by being privy to terrible and violent stories (told by both survivors and perpetrators of sexual violence) and with our seduction by these stories.

As Judith Butler so astutely pointed out in relation to the US response to the September 11th attacks, a consensus – arising out of hegemonic discourses – on what certain terms mean and 'what lines of solidarity are implicitly drawn through this use' emerges through the telling of familiar narratives, so that certain stories preclude the telling or hearing of other stories (Butler 2004a: 4). She explains that with the experiences of violence (such as the September 11th attacks) and, we would argue, also the collective (sudden?) (Žižek 2009) recognition of rape as a war crime and a critical and integral aspect of warring, a 'frame for understanding violence emerges' (Butler 2004a: 4).

> The frame works both to preclude certain kinds of questions, certain kinds of historical inquiries, and to function as a moral justification for retaliation. It seems crucial to attend to this frame, since it decides, in a forceful way, *what we can hear*, whether a view will be taken as explanation or as exoneration, whether we can hear the difference, and abide by it … (Ibid.: 4–5)[5]

In the main storyline of conflict-related rape as we know it (while including slight variations) the casting is clear: the role of villain/perpetrator is held by the man in uniform and the victim/survivor role is occupied by women, especially raped women. This overarching storyline, and the inhering subjects of these stories, is made intelligible through assumptions about gender (in an interplay with a myriad of other intersecting and mutually informing productive power relations, e.g. race, class, nation, etc.) (Hutchings 2007). As we will see, gender provides a framework for the drawing of many critical lines of distinction against which we can determine the just from the unjust, good from evil, ethical behaviour from unethical behaviour, victim from perpetrator, ordered lawful society from barbaric chaos, humans from beasts, the normal from the abnormal, as well as a familiar 'we' from a strange and terrible Other. Indeed, 'gender is ... often used to provide an ethical shorthand which helps to render certain kinds of positions of violence intelligible' (Hutchings 2007, 2008a). Furthermore, the ethical shorthand embedded in the stories we tell about rape and war in world politics enables us to act: to attend to the victims of violent acts, to protect them, to hear their voices, and to perhaps even attempt to heal them (see Chapter 4). These stories also allow us to identify (and punish/reform) the perpetrators and, therewith, even work to prevent further acts of violence.

The dominant story of wartime rape *feels* ancient and familiar. However, as we noted in the Introduction, both the main plot and the prominence of the problem of wartime rape in global policy forums have changed since the debacle of genocide in Rwanda and the exposure of 'rape camps' in the wars in the Balkans (Enloe 2000: 109, 134). The DRC, as we have explained, is perhaps the most infamous site of wartime-related sexual and gender-based violence, given its recently won (and highly dubious) status as 'the rape capital of the world' (Wallström 2011). Indeed, the 'Congo' can also be seen as ethical shorthand for signifying wartime rape and its attendant divisions of abject victim (Diken and Laustsen 2005) and bestial perpetrator in a context of barbarism and chaos, which is reliant on familiar colonial lexicons.

In the rest of this chapter we linger on the prevailing story of conflict-related sexual violence against the backdrop of how rape in the Congo is represented in governing global discourses. Although we discuss the Rape as a Weapon of War discourse as the dominant framing of wartime rape, we concentrate our analysis in this chapter on how a focus on sexual violence as gendered (instead of sexed) makes certain subjects both possible and impossible. More specific focus on the Rape as a Weapon of War discourse, which builds on this analysis, will follow in the next chapter.

We now turn to the story of sexual violence, which preceded, conditioned and haunts the current storyline of rape as gendered (and not 'sexed').

The 'Sexed' Story: biology, (hetero)sexual urge and substitution

The explanatory framework commonly understood as the 'Biological Urge/ Substitution Theory' casts rape as an natural but 'unfortunate by-product of war' (Niarchos 1995: 651). This explanatory framework precipitated hard political and academic work to establish the current notion of rape as a gendered weapon of war in its stead. The Biological Urge/Substitution Theory, with its own allocated subject positions and representations, nonetheless haunts the authority and certainty of the notion of sexual violence as 'gendered'. Moreover (and importantly for understanding how some – including many soldiers worldwide – understand wartime rape), this storyline still holds much purchase within the military, as well as within society at large (Enloe 2000; Wood 2009; Higate and Hopton 2005; Higate 2004; Whitworth 2004).[6] It nonetheless has become politically incorrect in most official policy arenas. Familiarity with this storyline, nonetheless, is important for being able to situate the current framings of sexual violence as a weapon of war in a wider repertoire of possible discourses. It is also important because the dominant grid of intelligibility for understanding the relations between sex/gender and violence and the subjects produced through (sexual) violence is formulated in both explicit and implicit relation to this other account. In order for the notion of sexual violence as 'gendered' to make sense, crucial claims or 'truths' (such as essentialist notions of male heterosexuality as a natural and formidable force that demands an outlet) inform the dominant framings through their very exclusion. Efforts to exorcize such claims or 'truths', however, necessarily fail because of their integral importance as points of contradistinction and because their traces underwrite the dominant plot of wartime rape. Such rejected notions are thus even inadvertently reinforced.

What, then, are the basic plots of the 'Sexed' Story? Simply put, historically, rape has been seen as integral to warring because war is (supposedly) enacted by men and men are subject to their biologically driven heterosexual needs; hence men rape. The main line of argument according to this explanatory framework is twofold.

First, the (male) soldier's libido is understood as a formidable natural force, which ultimately demands sexual satisfaction (ideally from women). Maintaining multiple sexual relations and displaying sexual potency are seen as 'natural' effects of male heterosexuality. According to this framework of understanding, often called the 'sexual urge' (Seifert 1996: 36) or the 'pressure cooker theory' (Seifert 1994: 55), wartime rape is a result of the heterosexual desires of men, resulting from their biological make-up (Paglia 1992; Thornhill and Palmer 2000).

This basic storyline comes in various forms, from the more determinist form in which rape functions to fulfil the male biological drive to perpetuate one's genes, to more biosocial theories (for a discussion of these variations, see Gottschall 2004). The more popular version of this 'sexual urge' discourse is often referred to as the 'substitution' argument (see Wood 2009: 135). According to this line of reasoning, sex by force occurs in military contexts because soldiers do not enjoy 'normal' access to women in other ways, as they are not granted leave, they are far from home, or owing to the basic travails of war. If men are not able to achieve sexual relief in the socially acceptable way (through consensual sex with wives, girlfriends or prostitutes), then they will 'substitute' sex by force for 'normal' sex out of sheer necessity. This is the familiar 'soldiers get horny and need an outlet' explanation, which easily glides into a 'boys will be boys' rationale. Many refer to the notion of a 'recreational rape' (or in the case of the DRC a 'lust' rape), which occurs if soldiers are deprived of the normal outlets for sexual desires (see Eriksson Baaz and Stern 2009; Enloe 2000: 111).

Following from this reasoning, in many military contexts men's sexual needs are often presented as the reason for the need for regular leave (also to reduce the risk of supposedly unhealthy homosexual acts) (Enloe 2000; Goldstein 2001). 'Solutions' to such sexual violence can therefore be found in increasing soldiers' access to women, either through more generous leave or through 'comfort women' (as was the case in Second World War Japan; see Chapter 3). Prostitution rings have surrounded military bases throughout history and in diverse global contexts, including UN peacekeeping missions (Higate 2004; Higate and Hopton 2005; Whitworth 2004). This line of reasoning is particularly dominant in military contexts, and underwrites generalized (military) accounts about male sexuality that celebrate virility and sexual potency.

Secondly, and intimately connected to the first line of reasoning, is the rationale that war suspends the social constraints that hinder men from being the sexual animals that they 'naturally' are/can be. According to this perspective, society 'normally' acts as a hindrance to males' natural sexual drives – a hindrance which is often removed in the climate of warring. As Stern and Zalewski have argued elsewhere (Stern and Zalewski 2009), this narrative reproduces the notion that boys are biologically and ontologically prior entities who will follow a certain predestined development into civilized citizen-men (also a known category) if given the right conditions. These conditions reside in their being in society, presumably a civilian space where they can be nurtured by mothers and later wives.

In this story, the army/military is a special domain, which is separate from the homeland – the sphere of civilian life where normal civilization resides.

Hence, when in a situation of extreme violence, all men, theoretically speaking, are potential rapists, as their biologically driven natural sexual urges are no longer tempered through society. Their 'natural' state as beasts is unleashed. This storyline gains purchase through its familiar resonance with established narratives about the nature of men in the 'state of nature',[7] most notably from Hobbes's rendition of social contract theory (Hobbes 1651; see Pateman and Shanley 1991, Carver 2008b for a discussion of Hobbes; see also Chapter 3).

It is important to highlight that many military staff (as well as people in general) understand conflict-related rape in this way. Indeed, in our research in the DRC, military personnel from the FARDC as well as external actors often described rape in war as somehow normal, as an unavoidable consequence of warring or as a consequence of bad discipline (Eriksson Baaz and Stern 2009) (this will be further discussed in Chapter 3) combined even with boredom (see note 6). This line of argument has been supported by much research in other contexts (Seifert 1996: 36). However, despite its prevalence as an accepted explanation of an unfortunate (yet unavoidable?) reality in military contexts and in society at large, official discourse, as well as academic arguments, refutes the Sexual Urge/Substitution Theory.

In sum, the 'Sexed' Story is organized around notions of male heterosexuality as a natural force. Gender is largely seen as inseparable from sex insofar as gender roles appear predestined, or at least prefigured through biology. In the context of warring (portrayed as similar to the 'state of nature'), the civilizing restraints of society are suspended. The subjects allotted through this discourse are then subordinate to the forces of nature: women appear as silent victims of the expression of men's biology, and men as subjected to the drives of their bodies.

The 'Gendered' Story: gender and militarization[8]

According to its critics, the above storyline is essentializing and deterministic as well as overly negative towards men as such. It also naturalizes and thus depoliticizes rape in war and waylays efforts to stop its occurrence. Building upon a wealth of feminist research into the connections between gender, militarization and warring (as well as the logics of security and national identity), scholars and, later, policy-makers/advocates instead shed light on the power of gender ideologies as underlying rationales for the 'use of' sexual violence in armed conflict. According to this explanatory framework, rape in conflict settings is seen as an effective tool of humiliation and intimidation. Many understand this as a vital component of rape as a strategy of war, as we will discuss further in the following chapter. Here we will mainly focus on how rape in war is made intelligible through gender.

Instead of seeing the military as a venue through which boys can achieve their natural potential as men, feminist research underscores how men/boys (and women/girls) learn to be 'masculine' and violent in the military through methods specifically designed to create soldiers who are able (and willing) to kill to protect the state/nation (see, e.g., Alison 2007; Bourke 1999, 2007; Connell 1995; Ehrenreich 1997; Enloe 1990, 2000, 2007; Goldstein 2001; Higate and Hopton 2005; Leatherman 2011; Morgan 1994; Pankhurst 2009; Pin-Fat and Stern 2005; Price 2001; Schott 2003; Shepherd 2007; Sjoberg and Gentry 2007; Stern and Nystrand 2006; Stern and Zalewski 2009; Whitehead 2002; Whitworth 2004). The logic of militarization, in part, depends upon particular articulations of ideal types of masculinity and femininity, whereby, through the discourses of war, men are cast as heterosexual masculine citizen-soldiers. By contrast, women (and 'the feminine') are stereotypically associated with a need for protection, with peacefulness and life-giving; these associations serve as the necessary counterpart to the supposed 'masculinity' of protecting, warring and killing (Enloe 1990; Goldstein 2001; Higate and Hopton 2005; Masters 2008; Pin-Fat and Stern 2005).

According to this line of reasoning, the desirable type of masculinity that is produced within the military celebrates violence, order, masculine-coded obedience and domination. It serves to form soldiers according to strictly disciplined codes of behaviour that designate any deviance from the norm as inferior, feminine, effeminate and dangerous. Boys/men undergo a form of indoctrination, which includes humiliation and breaking down of the civilian (feminized) boyish identity, and then the building up of the macho soldier. This occurs through, among other things, group bonding – even through the shared experience of group rape, which also fosters group loyalty (Alison 2007; Card 1996: 7; Connell 1995; see also Cohen 2011).

All that is associated with femininity is seen as corrosive of the required militarized masculinities. Therefore, violence is also directed inwards towards the '"others within"; killing the "women in them" becomes necessary for soldiers in their attempts to live up to the myths of militarized manhood' (Whitworth 2004: 176). In sum, militarization requires the production of different heterosexual violent masculinities (including both generals and foot soldiers); racial, ethnic and class hierarchies are 'woven into most military chains of command' (Enloe 2000: 152; see also Higate 2004; Higate and Hopton 2005).

Militarized (and mythologized) masculinities (and the attendant promises and entitlements associated with inhabiting these masculinities), however, rarely resonate with soldiers' sense of self and lived experiences, or with the actual conditions of militarized men's lives (Whitworth 2004: 166; Eriksson Baaz and Stern 2009). The fragility and indeed impossibility of militarized masculinity therefore requires continual concealment through military institutional practices, and in the individual expressions of such masculinity. This line of thinking builds upon a notion of identity as a continual process, which can never be fully realized; the fulfilment of any identity position is therefore impossible and bound to 'fail' (see Butler 1990; Hall 1996a). While 'inherently impossible', feelings of 'failed masculinity' can be seen to contribute to sexual violence in that rape becomes a way to try to perform and regain masculinity and power (see Eriksson Baaz and Stern 2009).

Importantly, the associations of women/girls with stereotypically 'feminine' attributes and men/boys with stereotypically 'masculine' attributes render women/girls particularly vulnerable to the logics of rape in conflict and post-conflict settings.[9] For instance, rape can be conceptualized symbolically as a way to punish, humiliate or torture seemingly 'subversive' women for threatening national security (and identity) through their perceived challenges to strictly defined notions of femininity and masculinity (Enloe 2000; Stern 2005: 86). The women in 'need of punishment' challenge the notion of femininity that is 'worthy of protecting'. As women are often cast as the symbolic bearers of ethno/national identity through their roles as biological, cultural and social reproducers of the community, rape of 'enemy' women can also aim at destroying the very fabric of society (e.g. Enloe 2000; Goldstein 2001; Stern and Nystrand 2006; Yuval-Davis 1997). Conceptualizations about ideal femininity, which link femininity to chastity and virginity, also play a particularly important role. These ideals add to the effectiveness of rape as an act of humiliation and destruction, since it 'sullies' women and renders them seemingly unsuitable for present or future marriage and love relationships. As we have seen in many contexts worldwide, wartime rape often results in the raped woman being rejected by her husband/family, or as rendered unsuitable for marriage. Moreover and importantly, rape in wartime can be seen as a particularly effective means to humiliate (feminize) enemy men by sullying 'his' women/nation/homeland, and proving him to be an inadequate protector (e.g. Enloe 2000; Goldstein 2001; Stern and Nystrand 2006; Yuval-Davis 1997).

In line with the above reasoning and with the long (and important) political struggle for sexual violence to be legally decreed a crime against humanity and a war crime, sexual violence in conflict settings is described and understood as a 'weapon' or 'tactic' of war in both academic circles and policy settings (see Chapter 2).

We now turn to a discussion of what the supposed move from sex to gender and the consequent singular focus on sexual violence as a particular form of gendered violence might inadvertently *do* to our understandings of the subjects produced through the available lexicons of sex-gender-violence.

Sex/gender?[10] In the 'Gendered' Story of sexual violence (in its necessary opposition to the 'Sexed' Story), the division between victims and perpetrators does not necessarily follow the division between female and male sexed bodies. Herein lies perhaps the most palpable difference between rape as sexed and rape as gendered. In the basic storyline of gender and militarization introduced above, the humans who commit acts of rape are gendered as masculine; masculinity emerges as a learned attribute. In most cases this means male sexed bodies, but can also mean female sexed bodies.[11] The victims of these acts of violence are gendered feminine, most often meaning female sexed bodies, but also including male sexed bodies that are symbolically feminized through the act of rape.

Gender, in this story, connotes masculinity or femininity, or the relations between the two. Focusing on sexual violence as 'gendered' implies focusing on how masculinity (and femininity) act on, impact, influence or provide roles for the sexed body, and therewith how gender works to underwrite or even produce the act of sexual violence. In the 'Gendered' Story, the focus is placed on socially produced masculinities instead of on sex/biology. Trouble resides in gender (as a learned attribute); not in the natural essence of man. This feminist story thus successfully shifts attention away from an essentialist reading of men's violent nature towards masculinity as a construction. We have seemingly left sex behind. Yet this story relies on our knowing the subject: men (as opposed to 'men' constructed through workings of gender). We seem to take for granted that we know who/what men are, and the problem lies instead in what types of masculinity (good peaceful ones, or violent militarized ones) these men will learn. Here we talk about men who are 'real' and pre-given subjects who are separate from, and victims of, the workings of gender. Furthermore, we claim that gender is a construction but treat it as not only real, but also as a diseased or afflicted accessory to the body that could be cured. The act that results from this affliction occurs *because* of gender. We promise that if only we could do gender (read: produce masculinities) differently, then the scourge of sexual violence might disappear.

Here we can glimpse how the performative production of sexgender works to ensnare us in a seeming impasse: the *sexgender paradox*.[12] Let us briefly explain. Our feminist representations of women (and men) do not correspond to some underlying truth of what woman/man is or can be; rather feminism produces the subject of woman/man, which it then subsequently comes to represent (Lloyd 2007: 26). Yet a common argument is that feminist scholarship performatively reproduces the sexed identities and attached gendered harms it sets out to eviscerate. Simply put, attention to gender as constructed nonetheless tethers gender to the sexed body that gender is seen to act upon.

The workings of gender thus rely upon 'sex' and the sex–gender distinction. Attention to the power of gender therewith also implies an implicit attention to and reproduction of 'sex'. We are nonetheless enticed into thinking that by paying attention to gender, we have refuted the power of sex. A grammar or logic of temporality glues feminism to a generational trajectory of activity always seemingly en route to proffering a solution to gender – unsurprisingly prefiguring an unavoidable failure to ever arrive. This failure occurs in part because there is a sense that when we speak about gender we leave the sexed body (as explanation for/cause of) gender behind, although at the same time our understandings of gender as a construction are premised by a distinct and meaningful category of 'sex' and sexed bodies upon which gender acts. We work to separate sex from *the construction of* gender and to concentrate on the power of gender, yet we fail fully to do so, as 'sex' haunts any rendition of 'gender' we may imagine.

This seemingly unfortunate but significant paradox, which has been well addressed in feminist theory,[13] has been nonetheless under-investigated in relation to gender in the military (cf. Higate and Henry 2004; Masters 2008; Stern and Zalewski 2009). As we will see, an understanding of the sexgender paradox (as it is briefly introduced here) aids in better comprehending the ways in which sex haunts gender in our reading of the 'Gendered' Story of militarized masculinities and wartime sexual violence, and how it plays out in the framing of wartime rape, as explored in the remainder of the book.

Available grids of intelligibility: sex-gender-violence in the DRC

Let us turn to the site of the DRC to flesh out what else might be going on in the plausible and politically vital storyline offered by the 'Gendered' Story and the inhering shift *from* sex *to* gender. The main story of sexual violence in reports on the DRC is that it is gendered – and, as such, also an integral part of general gender-based discrimination and subordination of women (Amnesty International 2008; Ertürk 2008; Human Rights Watch 2002; Kristof 2008; Ohambe et al. 2005; Wallström 2010b). Sexual violence is seen to be a gendered weapon of war in the DRC in the overarching and governing discourse, as we will see further in the next chapter. According to this discourse, this 'weapon' is made possible in part because of the unequal gendered relations that reign in society, and through the violent militarization of masculinities. The connection between unequal gender relations generally in society and the amount of conflict-related sexual violence is methodologically difficult to establish, given both the lack of reliable data and the difficulty of drawing a line between conflict-related violence and violence in society more generally.[14] This, however, does not seriously challenge the supposition that there is some sort of correlation between the two that attends this basic storyline (see Alison 2007).

Yet in the very insistence on the gendered-ness of sexual violence, the 'Sexed' Story is not left behind. The oft-repeated storyline in media as well as policy reports of violence in the Congolese warscape follows the familiar plot noted above. Female sexed bodies appear as *the* victims/survivors, with few exceptions. Sexualized and racialized depictions of Congolese armed men as particularly bestial, and violence as particularly chaotic and natural-ized – evoking familiar images out of the colonial lexicon, and proffering the 'Congo' as ethical shorthand for the scourge of wartime rape – pepper many accounts of rape in the DRC (Carroll 2005; Eriksson Baaz and Stern 2008; Gettleman 2007; Kahorha 2011; Guardian 2011; see also Chapter 4 for a further discussion). Sexual violence thus *also* emerges as explicitly sexed and raced; and the violence is depicted with a focus on the sexed body, often on brutal violence against women's genitalia.[15] Hence, while sexual violence is written as *gendered*, the performative predicament of sexgender lingers in the casting of perpetrators (men) and victims (women). Such lingerings thus disrupt the story of sexual violence as fundamentally – and politically necessarily – gendered.

Hence, the story of conflict-related rape in the DRC is relayed as if it were familiar and known, with its two seemingly contradictory, yet nonetheless mutually reinforcing, subplots: rape as sexed/raced and rape as gendered. Yet, importantly, despite its perennial (yet newly recognized as politically important) plot, rape in the warscape of the DRC is nonetheless rendered exceptional through both of these registers. What do we mean by its being rendered exceptional, and how might these renderings be important in our unpacking of the dominant grids of intelligibility for understanding sexual violence and its subjects? To better address these questions we now turn to a critical reading of the 'Gendered' Story of rape and rapists as it is employed to represent, frame, explain and even remedy rape in the Congolese warscape.

Sexual violence in the DRC as exceptional In the rush to attend to its harms, sexual violence in the DRC has been conceptualized as 'horrific' by its being specifically and sensationally gendered, sexualized and 'wrong' and fundamentally different from and outside of other forms of (legitimate?) violence. Indeed, as noted in the introduction, there has been a specific, often exclusive, focus on sexual violence in global reporting of the DRC conflict. The brutality and extent of rape that occurs there emerges as unparalleled, never seen before (Gettleman 2007; Human Rights Watch 2009a; Kippenberg 2009)! In such a context, 'vaginal destruction' is classified 'as a crime of combat'

(RHRC, 'Gender-based violence: key messages', cited in Horwood et al. 2007: 24), 'murderous madness' (Horwood et al. 2007: 15).

The logic underwriting the notion of the exceptionalism of rape in the DRC is again at least twofold: *first*, conflict-related rape (and rapists) in the DRC is/are necessarily exceptional to the seemingly civilized warfare and warriors that serve as its/their implied contrast in a teleological and racialized narrative reflecting the Enlightenment promise. *Secondly*, the seeming exceptionalism of rape relies on a more subtle move through which the very gendering of sexual violence (in explicit contrast to the 'sexed' story which serves both as its constitutive outside and as its parallel plot) renders sexual violence as *sui generis* – separate and outside of other types of violence. We briefly touch on the first line of argument, which is arguably quite straightforward, and then move on to unpack the more difficult twists and turns in the second. Importantly, through rendering sexual violence exceptional in different registers, those who rape emerge in the 'Gendered' Story as simultaneously normal (yet in need of reform) and abnormal, and even unhuman.

Let us now turn to the first logic to explore how rape and rapists are rendered exceptional through a teleological and racialized narrative.

Racialized bodies Through a single focus on what are described as bizarre and uncivilized methods of warfare, depictions, like those relayed in the opening quotations to this chapter, inform a generalized process of Othering. African wars in general, and the wars in the DRC in particular, appear in both the Western media and global policy reports as primitive, anarchic and barbaric, as fundamentally 'Other' (Broch-Due 2005; Chan 2011; Eriksson Baaz 2005; Gikandi 1996; Keen 2005; Pottier 2005). Soldiers and combatants emerge as brutal vengeful killers and rapists who mutilate and eat their victims (see also Broch-Due 2005: 33; Keen 2005: 3; Chapter 4). We briefly introduce the notion of the intersections of race and gender[16] here in order to pick through the sexgender paradox as it plays out in the story of rape and rapists in the DRC.

The acts of rape that have occurred in the DRC are often represented as a result of the supposed bestiality of the rapists. While rarely directly stated, this is intimated through not so subtle allusions (e.g. Guardian 2011). As we will further discuss in Chapter 4, sexual violence thus also emerges as explicitly sexed and raced. In this sense, rape, violence and the perpetrators of these acts in the DRC emerge as exceptional to modern warring. The idea conveyed is that modernity has enabled civilized peoples to abandon such bestial practices and to abide by the laws and norms of warfare. That this assumption has poor anchoring in the experiences of warfare in the supposedly civilized world seems to have little bearing on its purchase.

Colonial and racialized scripting of the conflict in the DRC and its main players portrays them as continuing to reside in a bygone era where and when beasts rule the jungles. Despite their simplistic, racist and mistaken base, such portrayals seem to offer a reasonable lens through which to address conflict-related rape in this context. This reasonableness resides in the implied notion that this backward state can be rectified through, for example, the enlightenment and modernization of both the armed forces and society more generally. The Congolese too can thus leave behind the dictates of bestial sexuality and learn to be more productive and less violent gendered subjects.[17]

In this narrative, the Congolese rapists emerge as '*normal*' in the sense that they are governed by their 'natural' (barbaric) essence and hence follow the norms determined by nature; yet they are abnormal in the sense that they deviate from the norms of modern 'civilization' (Dunn 2003; Mudimbe 1994). The sexgender paradox thus receives a temporal solution: the power of sex is the teleological precedent of the power of gender.[18] Images of the barbaric African (masculine) Other, who is unleashed by the conditions of war to act according to his 'true' nature, both complement and disrupt the parallel storyline of rape as a gendered weapon of war. They also complement and disrupt the story of rapists as embodying (or adorning) a violent masculinity constructed in the military and in unequal gendered power relations in society more generally (Eriksson Baaz and Stern 2008).[19] The raced/sexed story complements the gendered story through its anchoring of sexgender on to specific kinds of bodies – racialized bodies that are necessarily Other. These bodies are Other because of their backward reflection of an uncivilized site, which was seemingly left behind, through, among other things, a revamping and enlightened *modernization* of gender.

The sexgender paradox is thus seemingly smoothed over through race, insofar as certain 'backward' racialized bodies are mired in 'sex', while civilized modern bodies are free of sex and subject to different configurations of gender. Through this equation, sex precedes gender in a teleological narrative. Gender makes sense through its exclusion of, and reference to, the sexed body that is not 'ours', but belongs to the Other. This 'sexed' body serves not only as a constitutive outside to the gendered body, but as a reminder that the confines of the past can grasp one in their retrograde clutches once again. The comfort of the gendered story as a mode of understanding performs coherently. Yet the sex story interrupts the main storyline of sexual violence as fundamentally gendered. It does so through its insistent focus on the body and the consistency of 'sex' in terms of the casting of comprehensible and not too messy categories of victims (female) and perpetrators (heterosexual males). Gender is thus rendered merely a consequence (albeit an avoidable one, perhaps) of sex.

Gendering sexual violence – revisited Yet if we further focus our critical gaze on the gendered story (while of course keeping in mind how this story is haunted by the racialized 'Sexed' Story), what else can we see? First we explore how gendering sexual violence discursively produces 'normal' and, importantly, reformable subjects in different, even contradictory, ways. We then discuss how the singular focus on sexual violence discursively casts the 'rapist' as abnormal and ultimately dehumanized.

Feminist insights have taught us that violent militarized heterosexual masculinity must be carefully *produced* through militarization more generally and military training specifically (Braudy 2003; Connell 2000; Enloe 1990; Goldstein 2001; Higate and Hopton 2005; Hutchings 2008b; Parpart 2008; Pin-Fat and Stern 2005; Woodward and Winter 2007). Moreover, the process of becoming a citizen-soldier who is capable of protecting the nation/state/group and killing the enemy persists through the practice of warring; the production of militarized masculinity is a continual process, which never fully 'arrives'. According to the familiar feminist fable (Stern and Zalewski 2009), the production of such masculinity is required by the gendered logics of warring (which depend on the gendered associations of the masculine with killing and violence, and the feminine with life and peacefulness). Hence, sexual violence performs as the fault of gender in the sense that it is an expression of the violent masculinity necessarily produced through the military. Simply put, men are *produced* as rapists through militarization. In this sense, soldiers who engage in sexual violence are answering the call[20] to fulfil ideal types of military masculinity and in so doing are adhering to established norms (which may or may not be sanctioned though specific orders and military justice systems).

Furthermore, as we will see in the following chapter, conceptualizations of sexual violence as a weapon of war make sense because rape is understood as a *gendered* act, effective *precisely because it is gendered*. According to this logic, rape is a violent act not only against female sexed bodies, but against the 'enemy' as such through the logics of gender. (In the case of the Congo, while often evoked as a blanket explanation for the occurrence of sexual violence, such generalized assumptions about militarized masculinity and about rape as strategic poorly reflect the complex web of relations, motives and experiences on the ground, as we will see in Chapter 3.) Nonetheless, in the generalized gendered storyline summarized above, rape as a weapon in wartime is made possible through linkages between particular constructions of masculinity, femininity and violence/war/peace/public/private/home. *Understood this way, those who rape in warfare are the 'normal' subjects successfully produced through (violent and nefarious) gendered militarization.* Again we see how the 'rapists' emerge as subjects in need of reform, except here the fault lies not in the rapists themselves primarily, but in their success at inhabiting a violent sub-ject position. Those who are raped, similarly, emerge as hindered, delimited, even maimed by their successful production as victims (and symbols of group belonging) by gendered power relations (see Stern and Zalewski 2009 for further discussion of this). A way to redress the known outcome of such successful gendering is to gender differently (e.g. produce less violent, more 'civilized' military masculinities, and more active femininities), or relatedly, to alter the contours of militarization so that different masculinities and femininities are required.[21]

Another subplot to this story is the flipside of successful gendering: namely the male soldiers' failure to inhabit and enact characteristics of ideal militarized masculinities as an impetus to rape. Following this line of thinking, soldiers who rape do so in part out of the frustration, disappointment, anger, anxiety, etc., that arise in the acute discord between their embodied experiences and their expectations of themselves as soldiers (men) in the armed forces (see also Higate and Hopton 2005; Eriksson Baaz and Stern 2009). While successfully inhabiting any idealized subject position is impossible, the strict dictates of militarized masculinities accentuate the arguably 'normal' sense of failure and the frustration that attends it. This sense of failure is 'normal' in that successfully inhabiting subject positions is integral to identity formation gen-erally. Rape serves as a performative act that functions to reconstitute their masculinity – yet simultaneously symbolizes their ultimate inability to do so. In the site of soldiering, the act of rape thus constitutes a double failure. First, it is conditioned by the failure to act in a sexually 'normal' way, and secondly, it symbolizes the failure of the performative effort to become masculine in the context of the 'abnormality' of war (which suspends the social mores and gender norms of society) and poverty. In sum, their 'failure' to be 'normal', as defined by their expectations of themselves as masculine soldiers, precipitates the violence soldiers who rape enact, in part, in their efforts to *be* normal (see Eriksson Baaz and Stern 2009; Whitworth 2004). Ironically, these very normal failings (in the sense that all identity processes are impossible) to be 'normal' (understood as the ideally fulfilled subject positions of militarized masculinity) in the context of warring produce highly violent effects.

The possible subjects crafted through this subplot are bound by its grid of intelligibility. Both those who rape and those who are raped emerge as the victims of the impossibility of gender – in particular, militarized masculinity. According to this logic, the allure of impossible masculinity in the midst of extreme conditions of warring, and without the many possible subject positions of 'normal' society, traps men in an insufferable position, which they seek to alleviate through the enactment of rape. Here we see again how gender emerges as an affliction imposed upon sexed bodies, whose constraints harm both those who rape and, of course, those who are raped.

In the above paragraphs we have explored how the gendering of sexual violence (and the apparitions of its sexed parallel plot) have rendered the subjects of sexual violence 'normal' – and yet potentially reformable through their very gendering. The point of this exercise is to make visible the available grids of intelligibility through which we understand sexual violence and its subjects, in this case in the context of the DRC warscape. We now return to the phenomenon of the singular focus on sexual violence in the DRC and scrutinize the relationship between sex/gender and violence. We do so in order to glimpse another aspect of the work gender does in establishing an *ethical shorthand* for making sense of, producing and remedying violence and its subjects.

As noted above, in the rush to attend to its harms, sexual violence in the DRC has been conceptualized as 'abnormal' by its being specifically and sensationally gendered, sexualized and 'wrong', and fundamentally different from and outside of other forms of (legitimate?) violence. It is in this rush that the 'normalcy' of rape and rapists discussed above gets trampled by the need to clearly discern victims from perpetrators, good from evil, and humans from beasts. The possibility of reforming those who rape lingers nonetheless in external interventions for defence reform (see Eriksson Baaz and Stern 2010) without yet seriously imperilling these lines of distinction.

The singular focus on sexual violence in the case of the Congo as exceptional and importantly as gendered, animates gender (and the 'Congo') as ethical shorthand. This ethical shorthand keeps beasts and even violence separate from or outside of the self, of humanity and of 'the human'.[22] In general terms, the masculinity that occasions rape performs as a hypermasculinity (Parpart 2010; Munn 2008: 153–7) in (at least) three different registers: 1) masculinity reminiscent of barbaric times (through the racialized tropes noted above); 2) masculinity devoid of humanity (along the lines of Masters's notions of cyborg soldiers (Masters 2008, 2010, 2012; Carver 2008a) as a machine of war); or 3) sick, riddled with the cancer of violence and evil and/or the frustration erupting through its inevitable failings. The overall picture is one of gender gone awry and monsters being produced instead of men.

This is particularly the case in representations of militarized masculinities in the DRC. If we return, for example, to one of the opening quotations of this chapter, we can recognize the production of non-human beasts in the descriptions of the rapists appearing out of the forest:

> They came out of the forest. Men with guns appearing barely human to the frail, ageing woman who months later recounted her ordeal, bent double after surgery to save her womb. 'They didn't look like men. Their skin was covered in cuts. Their clothes were completely torn. They became someone else, not humans.' (McGreal 2008)

This shift from (constructions of) men to that of monsters or beasts is slightly different from the move in the 'Sexed' Story by which 'men' slip backwards in progressive time to the biology they have left behind. Instead, the slippage occurs through a subtle circumscription of what being human means in contradistinction to the beast. Dehumanization occurs instead as the rapist's masculinity is cast as deviant to and outside of recognizable gendered norms. A sharp line of distinction between the rapist and the rest of humanity casts the rapist as fundamentally other. Sexual violence, and he who commits it (for in the gendered storyline, the rapist is gendered 'he', irrespective of sex), appears foreign, other, unethical and ultimately unhuman in relation to our socialized human selves. If we accept that the human is recognizable through its distinction from the beast/monster (Carver 2008a; Pin-Fat forthcoming), that is as decidedly social and so gendered, then the activity which does not fit in with these ways of being – indeed which threatens the very category of the human – must be exorcized. The wrongly gendered is then expunged from the realm of the human as deviant, as unhuman. In this sense the human stands as the *correctly* gendered human.

Rape, gender, humanity? Yet another slippage looms in the 'Gendered' Story, however, which serves to further the distance between the occurrence of sexual violence in the DRC and the realm of the human, but which leaves no discursive space for humans at all. Rape is written as a particular form of violence, emerging from its *gendering*.[23] As we discussed above, rape emerges as a fault and a product of *gender* – albeit deviant gender. Humanity is at once defined via socially 'correct' gendering (among other power relations) *and* as somehow above, beyond, a priori to the workings of gender (which arguably work upon the human), which resides somehow outside of its gendered manifestations. (This is, of course, the sexgender paradox revisited.)[24]

Understood in this way, gender arguably stands in between being human and violence, which is committed by non-humans, by default/definition. Let us further explain. Taking into account that we assume that the violence that others commit is unrecognizable as human activity, we can better see how gender can act as a borderline between humanity and violence. As we saw above, according to the 'Gendered' Story, (human) men become 'masculine' through the workings of gender. The form of military masculinity that is produced transforms them from men into *violent* beasts/machines of war (or as deficiently human through being sick/evil, depending on which version of this story is at play).

Yet this narrative of sexual violence prompts us to forget that it has also instructed us that humanity is *already* gendered. Instead, it separates out the wrongly gendered from the human, which somehow emerges as momentarily ungendered, and places the blame for the sexual violence that we abhor on 'gender' (that is, gender understood as something that acts upon the human subject). Importantly, this move renders this subject non-human, monstrous. As Elshtain (1987), Enloe (1983, 1990, 2000, 2002) and Hutchings (2008b) among others have shown us, being human in the lexicon of warring *means* being gendered in *specific* ways. Hence this storyline seems to fold in on itself and the sexgender paradox leads us down a path of the human being repelled by violence. This elision occurs because the spectre of the racialized 'Sexed' Story renders sexual violence as violence on the sexed body; the idea that all violence is gendered quickly evaporates.

We *do* recognize the human, however, because, of course, the human resides within us, who are decidedly not rapists. If, within this grid of intelligibility, we cannot accept sexual violence (a particular form of violence in the wide repertoire of violences) as belonging within and to the self/human, then this violence itself is *dehumanized*, understood in the sense that (wrongly) gendered humans ('males' and 'females' and masculinized and feminized bodies) commit violence. Gender allows for a distancing and ultimately an abrogation of responsibility for rape: if it is not a human activity, then it is not within, or of, us. There is no discursive place in this narrative for 'humans' who commit sexual violence.

Additionally, and especially important in the context of the Congo, the specific, often exclusive, focus on sexual violence hampers our understanding of the relationship between sexual violence and other (supposedly) ungendered violence. The 'Gendered' Story's important lesson that all violence is gendered is thus forgotten, and the blurriness between sexual violence and other forms of violence is rendered distinct. Furthermore, its emphasis on sexual violence as gendered and *sui generis*, of itself, discursively sets the stage for ultimately rendering 'other' violence in warring as 'normal', even naturalized. While this is not exclusive to the DRC, it is particularly evident in a context where other forms of violence are committed on a large scale against civilians, but largely and comparably receive much less attention.

In sum, as we have discussed above, gender and violence (as imbued with meaning in the 'Gendered' Story – and the 'Sexed' Story which haunts it) are co-productive. As Laura Shepherd explains, violence performs an ordering function in the reproduction of gendered subjects; and gender in the reproductions of the subject of violence (Shepherd 2007: 250). In our urgency to attend to the 'murderous madness' of sexual violence (Horwood et al. 2007: 15), we seek to render this madness recognizable in order to comprehend it and to act (see Chapter 4). By casting those who commit rape as bestial/ monstrous, and therewith decidedly not 'us', we miss the complexities and interconnections between subjects and the wide repertoire of violences that occur in the Congolese warscape. Hence, the violence of rape is constrained to a frame of understanding that on the one hand casts it as 'normal', but foreign to society and therefore requiring to be cast out of society; and foreign to ourselves, but created by society – yet also requiring expelling from society because it does not belong. The other (within and with) us is thus annihilated (Pin-Fat forthcoming).

Uncomfortable subjects

As we have seen, certain subjects are rendered possible and knowable via the available and dominant grids of intelligibility for understanding sexual violence. These grids, however, provide an impoverished framework for seeing, hearing, making sense of, writing about and empathizing with (Sylvester 1994) subjects of sexual violence. In the following paragraphs, we shift our voice from the difficult register of analysis and briefly touch upon some uncomfortable subjects, who/which do not neatly fit into these frameworks. We briefly discuss 'other violence', men/boys as survivors and women and girls as perpetrators of sexual violence in war, and the 'rapist' as a suffering subject.

'Other violence' It might seem strange to write of 'other violence' as an uncomfortable subject, when conducting research in a warscape that is notorious for its widespread, diffuse, intense and long-standing violences. Nonetheless, as is further developed in Chapter 4, the occurrence of other types of violence and, importantly, the complex relationship between different violences emerge as side stories (if present at all) in the rush to make sense of the scourge of war-related rape in the DRC. The subsuming of the complexity of violence into a neat narrative storyline is especially remarkable in the DRC context, where other kinds of violence against civilians are committed on a large scale – but left uncommented on, silenced (see Eriksson Baaz and Stern 2010). Many testimonies of rape that are relayed in reports, for instance, feature men, women and children being killed or mutilated. We offer two examples to highlight a mode of reporting which reflects dominant ways of perceiving (and not perceiving) violences. Our point is not to discredit the valuable contribution they make in raising awareness about conflict-related sexual violence, but to query what this awareness obfuscates. For instance, in one of the few testimonies provided in the 2008 report[25] of the UN Special Rapporteur on violence against women, one can read:

One woman from Nindja Village described how she and other villagers fled their homes every night to seek protection from the bush. One day in 2005, a group of Kinyarwanda-speaking men, whom she described as Interahamwe, tracked them down. They first executed the leader of the villagers and later the woman's brother, when he refused to rape her. The woman, her two children and about 50 others were marched off into the forest. When one of her children fell, the perpetrator forced her to kill her child. In captivity she was raped by 19 different men. On one occasion, the commander of the group forced her to fry and eat the hands and feet of her murdered sisters-in-law. At regular intervals, the militia would execute abducted women and men, after forcing them to first dig their graves. (Ertürk 2008: 8)

In one of the DRC testimonies in the OCHA/IRIN report *The Shame of War* we find the following account:

A few moments after the Interahamwe arrived in the village, I heard my neighbour screaming. I looked out of the window and I saw some men, all holding rifles. Immediately, I wanted to run away and hide but three of them turned up at our house. My husband pretended to be asleep ... they grabbed me roughly. One of them restrained me while another took my pili pipi pestle and pushed it several times into my vagina, as if he was pounding. This agony seemed to be a never ending hell [...] then they suddenly left. For two weeks my vagina was discharging. I was operated on ... I have to relieve myself into a bag tied to an opening in the side of my belly. They also killed my husband and my son. (Horwood et al. 2007: 15)

In both these testimonies, men, women and children are killed. In the first story, a man is killed because he refused to rape. Moreover, the woman raped is also forced to kill one of her children. These two violent acts, however, appear as side plots in the report – side crimes – to the main plot: that of gendered sexual violence and the nineteen times the woman was raped. In the report, the multiple harms expressed in these personal stories are effectively reduced to a story of rape (heard through the registers discussed above). Importantly, they are not further commented on, as the report goes on to discuss sexual violence (which this story is presented to exemplify) but does not address other violences or how these violences are interrelated. In the second story, a woman's vagina was destroyed: the main story. That her husband and son were killed at the same time is mentioned in passing – and, importantly, not commented on further. While the main story does indeed depict egregious harms, they are certainly not the only harms inflicted. How can it be that the story of rape need not include how rape is part of and relates to other violences? Why is rape (even if it is committed unimaginably and horrifically nineteen times) the main plot, while being forced to kill one's child becomes a side story? How can we hear these sufferings and not end up in an impossible situation, alluded to in the previous question, where the venue for talking about these 'other' violences is through a competition as to which violence is more gruesome and harmful, and not as relational? We will return to a further discussion of the difficulty of hearing and listening to other violences in Chapter 4.

Men and boys as victims/(non-)survivors; women and girls as perpetrators of sexual violence in war As discussed above, in the often repeated stories of sexual violence in the DRC, female sexed bodies appear predominantly as victims/survivors. While women and girls undoubtedly are the main (non-)survivors of rape, men also are victims of sexual violence in conflict settings (Carpenter 2006; Dolan 2011; Eriksson Baaz and Stern 2010; Johnson et al. 2010; Lewis 2009; Sivakumaran 2007, 2008).[26] As with sexual violence against women, sexual violence against men and boys in conflict is often quite varied, going beyond rape.[27] Carpenter (2006), for instance, has identified three main types of sexual violence experienced by men in conflict situations: 1) rape and sexual mutilation; 2) civilian men and boys being forced to rape; and 3) 'secondary victimization', in which rape of women forms part of a psychological torture against men. Testimonies of rape in the DRC draw attention to all these aspects of sexual violence experienced by men: male combatants being forced to rape; civilian men and boys being forced to have sexual intercourse with kin (daughters, mothers, wives) publicly (if they refuse, they have then been punished, often by death); and men and boys subjected to other violent and denigrating sexual acts (e.g. being dragged with a cord connected to the penis or the testicles, etc.).[28] However, men and boys have largely been rendered invisible as victims of violence in both research and policy papers by international organizations.[29] As Lewis (2009) has shown, international legal instruments have been developed in a way that often excludes men as a class of (non-)survivors of sexual violence in armed conflict (see also Sivakumaran 2010).[30]

This invisibility reflects a general neglect of the ways in which wartime gendered violence also affects civilian men and boys. While the dominant story of war throughout history most certainly has been that of men on the battlefields, the more recent and welcome attention to women in conflict settings (most often conducted under the rubric of conflict-related gender-based violence) has tended to forget the ways in which civilian men and boys are targeted in conflict settings. While women undoubtedly are the main (non-) survivors of rape in conflicts, men tend to be more vulnerable to 'other' conflict-related violence, in particular forced recruitment into armed groups, mass killings and forced labour (Jones 2000, 2002; Carpenter 2006). As Jones and Carpenter argue, since men/boys are particularly targeted here *because* they are men/boys (and are expected to fulfil common 'masculine' roles, such as soldier, combatant, etc.) this violence is also gendered. However, as Jones (2002: 76) concludes (in relation to the genocide in Rwanda), 'the general trend in discussions of "gender" and "human rights" tend to take women's disproportionate victimization as a guiding assumption, indeed almost as an article of faith'.

Another uncomfortable subject emerging from the 'Gendered' Story is that of women and girls as perpetrators of violence. While women and girls have participated in violent acts (including sexual violence) in the DRC as members of the state armed forces (FARDC) as well as the various armed groups, both the reporting of their acts and their voices have been largely absent in academic and policy debates, as well as in the media.[31] In line with global norms, a 'collective amnesia' (DeGroot 2000) seems to surround women's contributions (and complicity in violence) as combatants. While surely not unique globally, the silence surrounding female combatants has been particularly evident in the DRC context. In contrast to many other post-conflict contexts in Africa and elsewhere, the question of women combatants (as well as women's representation in security forces) in the DRC has been glaringly absent from the policy agenda.[32] This silence, we aver, must be understood in relation to the current singular focus on sexual violence in the DRC context, with its rush to attend to the victims of the scourge of rape. Messy categories of victims/perpetrators trouble existing mechanisms for redressing sexual violence.[33] Furthermore, the dominant gendered storyline recounted above provides us with a convincing and politically viable framework for making sense of women soldiers as masculinized through the workings of violent masculinity.

The occurrence of sexual violence that troubles accepted sex/gender norms (both women as perpetrators of sexual violence and men as survivors of sexual violence) is indeed slowly but surely being recognized in policy reports and even in the media (Gettleman 2009a; Storr 2011; UN OCHA 2008). However, we address this here as an 'uncomfortable' subject because in our attempts to make sense of those few references to male victims of sexual violence (and to female perpetrators; see Eriksson Baaz and Stern 2013a) we found ourselves continually pulled into the received framework of the 'Gendered' Story. In this framework, these subjects of violence and their experiences emerge as feminized exceptions: men as victims/survivors are reduced to exceptions to the (implied) real victims of sexual violence: women and girls; women as perpetrators emerge as so masculinized that they no longer fit into the notion of 'woman', or appear as masculinized deviant monsters (see Bourke 1999; Eriksson Baaz and Stern 2013a; MacKenzie 2010; Sjoberg and Gentry 2007). These uncomfortable subjects' experiences and suffering (and desires) are made intelligible to us through the familiar binaries – men/perpetrator and women/victim, sex/gender – that underwrite the notion that sexual violence concerns and involves only female sexed bodies, and thwarts recognizing the agency of women/girls.[34] We are hindered from recognizing them in a different and more open register, one that does not primarily cast them as an exception or an example of an already known harm.

Perpetrators' stories If we return to the opening quotation and the subsequent questions that framed this chapter, we see that the uncomfortable topic of the subject who rapes has been the catalyst and the point of departure for the journey this chapter has taken. We will revisit the notion of the rapists' story in order to reflect on the discomfort their stories conjured within us.

During the course of conducting research on violence and the FARDC, we were frequently asked the same question when talking to people (mostly people from the so-called international community) – a haunting question. 'How is it? It must have been very difficult to conduct those interviews and to speak to those people?' (read: awful/bestial rapists). Most of the time we braced ourselves and responded: yes, it was (and is), but our sympathies were elicited through listening to the soldiers' own stories, not primarily or only – as one would expect – through listening to the acts that some of them had committed and/or were defending. The explanatory framework with its ready-made ethical shorthand became visible through its contrast with the glimpses we ascertained of another story – the 'perpetrator's' story. Our feelings did not find a 'proper' place in the available imaginaries of sexual violence and its subjects.[35] The gap between the often extreme dehumanization of the rapist/soldier in the story of rape in the DRC and the encounter with these soldiers sometimes felt unbridgeable. In their words, some of those we spoke to had been implicated in sexual violence not only as rapists and/or supporters of rape but also as forced spectators of family members being raped, maimed and killed.

This other story (at least the story we thought we could hear amid the clamour of familiar resonances) was about how the soldiers made sense of rape and violence, and their suffering both as perpetrators of this violence, and more generally. As evidenced in the second opening quote above, the soldiers spoke of their suffering, desperation, disappointments, poverty; of humiliation, of the violences they have endured – for many as child soldiers, but also in the everyday violence of the warscape in which they live and work. They also spoke of their struggles to provide for their families, of witnessing the death of their children, and fearing their own deaths; of their devotion to God and to their families. Their stories, sometimes violent and repellent to us, revealed the normalcy of the violence they and their colleagues committed, and the (necessary?) distance they felt towards the victims of their acts. They also provided us with a glimpse of how they reflect on what this violence has done to them, as men, husbands, soldiers. In sum, their stories were undoubtedly about human suffering. Importantly, they were also about struggles and choices around questions of ethics and their sense of themselves as agents and therewith ethically responsible for the rapes they commit (see Eriksson Baaz and Stern 2009).

Nonetheless, our compassion for their anguish was accompanied by guilt, unease and doubt. Several troubling and uncomfortable questions remain: How do we write of the subject who commits rape when the dominant discourse forcibly denies his/her humanity? As Elizabeth Dauphinée suggests, it may be that in order for us to separate ourselves fully from the violence that the perpetrator commits, we need to cast 'him' as inhuman:

[t]he inherent humanity of the perpetrator betrays our desire to dehumanize him in the face of his dehumanizing actions. Our attempt to explain, understand, and memorialize the dehumanized victim can also be understood as an attempt to bear the suffering that lies at the core of dehumanization. In our identification of the victim as a victim of inhumanity, however, the victim's dignity can somehow be restored through the stripping of the perpetrator, through the distancing as far as possible of the Self from the perpetrator. In so doing, we abandon the perpetrator. The perpetrator becomes Other. It is not me who injures, who betrays, who kills, who identifies those who will become victims. I have washed my hands. (Dauphinée 2007: 119)

If we follow Dauphinée's reasoning, and learn from our above discussions, then how can we sympathize, empathize, with the 'perpetrators' and name them also as 'victims', that is crucially as *also human*? What might doing so entail for us and for those who were subjected to their violent acts? How can we cooperate with them empathetically (Sylvester 1994)? Without pretending to offer a tidy solution, we linger over these dilemmas in our concluding discussion.

Conclusions: rendering the lives of rapists 'grievable'

> It is precisely this which leads us to understand the frenzied need to frame the disaster in comprehensible terms – the need to reconcile the disaster with the knowledgeable, the calculable, with that which can be analyzed and overcome. The Hague, the trials at Nuremburg and in Arusha evince the need to arrange events into a historical narrative that can be thus exceptionalized so that ethics does not founder on the shore of absolute ruin. It also poses a binary in the assignation of 'good' and 'evil' associated with specific groups of actors. It requires victims and perpetrators, prisons, prosecutors, spectators who can serve as witnesses and give testimony. These categories cannot be too messy, lest the story become so confused that it can no longer be told in a comprehensible way. One might be, yet cannot be, relegated to silence. (Dauphinée 2007: 86)

To conclude this chapter, we grapple with the questions (re)posed above and also return to some of our initial questions inspired by Judith Butler's work. How do we recognize and relate to the face and the voice of those who commit rape so that we can differently ask how rape and rapists become possible? And, how do we handle the dilemmas that doing so pose for us? Attending to these questions entails reflecting on our own complicity in terms of reproducing the familiar subjects of violence in different ways.

One pressing query which has accompanied our efforts to write about sexual violence and rapists differently has propelled us into uncomfortable sites, positions and self-reflections: simply put, have we done harm to women and feminism more generally?[36] More specifically, by writing about rape differently (however slightly), have we inadvertently harmed both present and future victims of sexual violence, as well as the hard-won purchase that global attention to conflict-related violence against women/girls has achieved? Through, for example, our attention to soldiers' stories (Eriksson Baaz and Stern 2008, 2009, 2010, 2011), did we become complicit in efforts towards halting, or even revoking, the much-needed focus on women and particular understandings of the workings of gender that, for example, UN Resolution 1325 (2000) and the OCHA report *The Shame of War: Sexual violence against women and girls in conflict* (Horwood et al. 2007) exemplify?

We cannot, however, answer these and other questions about the ethics of our analysis in any ultimately satisfactory way, and we return to them in different ways throughout the remainder of this book. Nonetheless, our response to this query here is twofold: 1) we, as researchers, can never know what effect our research has, if any, on individual lives, or in shaping political imaginaries. This, of course, does not exempt us from ethical responsibility. We believe, however, that disrupting the confines of certain frameworks of understanding, by attempting to hear other voices and make sense of other subject positions, *may* ultimately strengthen attempts at redressing the harms of sexual violence by expanding our grids of intelligibility.[37] 2) The implied notion that to comprehend (in however limited a way) horrible acts and to feel compassion for those who commit violence involves also condoning this violence invites us to further reflect on the ethical dilemmas and fears that conducting and partaking of such research elicit.

We return to Butler's words which opened this chapter to further reflect on how it may be possible to empathize with the suffering of those who rape, rendering 'him' human and recognizing our intersubjectivity, without ultimately being complicit in 'his' violence. Why complicit? If we accept that 'we' emerge in relation to an other, that being is 'always a being-with' (Edkins 2005b: 379–80), then we are faced with the other (the rapist, the perpetrator) as part of 'us' (Butler 2005). Butler explains as follows:

> [violence] delineates a physical vulnerability from which we cannot slip away, which we cannot finally resolve in the name of the subject, but which can provide a way to understand that none of us is fully bounded, utterly separate, but rather that we are in our skins, given over, in each other's hands, at each other's mercy. This is a situation we do not choose. It forms the horizon of choice, and it grounds our responsibility. (Ibid.: 101).

If we recognize the humanity and suffering of those who rape, what then separates 'him' from us: are we more afraid of the 'rapist' or the 'human' who rapes, because 'he' could also be us? Similar questions of responsibility and complicity have surely troubled many who seek to hear the voices of those considered 'evil', 'deviant' or 'deranged'.[38] Hence ... how do we attempt to understand the suffering of those who rape (offered as 'causes' of the context of rape, if not of the actual rapes themselves) without condoning their violent acts. We find ourselves in quite a quandary, as the distinctions between victim/perpetrator, normal/abnormal blur and, ultimately, the distinctions between I ('normal' researcher) and 'you' ('abnormal' rapist) come into question.

In this chapter we have attempted to pick through the grids of intelligibility for understanding sexual violence and its subjects. Sexgender works in simultaneously rendering those who rape normal and abnormal – but ultimately dehumanizing them. Instead of it being a progressive vector for understanding and even producing subjects of violence differently, we have revealed how sexgender ultimately works in these politically progressive storylines to separate us from the rapist and to dehumanize. This has made it difficult to listen to, and write of, sexual violence and its subjects differently. What, then, are we to do?

If we lose the distinctions above, then we also lose the ethical distinctions (which were written though the lexicons of gender and 'the Congo') between the victim/perpetrator, good/evil. We are afraid of being complicit by rendering the rapist's life and suffering 'grievable', by being in solidarity with him (Jabri 2007a), by rendering him human (or recognizing his humanity). As Skjelsbaek has also noted, 'Few have wanted to look into the mechanisms that create perpetrators, perhaps out of the fear that the possibility of committing the same crime is a potential we all have' (Skjelsbaek 2001: 212).

If these interconnections become undone, as they must, we will be forced to recognize the other; forced to realize that we are ethical subjects only in relation to the other. Our subjectivity entails being with another. Hence we exist in an intersubjective relationship with the other (Butler 2005; Jabri 2007a; Vaughan-Williams 2007). I and you blur; and this is the really scary part: I could become you, as there is nothing left that is seemingly solid which separates us: it could have been me, 'subjected to force' (Weil 1965). Maja Zehfuss explains this as follows:

> Our violence towards others, even in the name of defending ourselves or liberating them, affects who we are because we become who we are when we act. We are vulnerable in the most profound way imaginable not merely because we are physically vulnerable to others but because we may never know just who we are: we will only ever have been in our relation with others. (Zehfuss 2007)

Gender as it is produced through the sexgender paradox in the 'Gendered' Story of sexual violence seemingly rescues us from this precarious position. It provides an ethical map of distinction between self and Other, and enables us to abrogate from recognizing ourselves in (those) others. Gender thus seemingly comfortably stands between 'us' as human and those who rape. If 'he' (the rapist) is within the range of humanity – ungendered in the sense of momentarily broken free[39] from his prescribed racialized and gendered self (which works to designate the normal from the abnormal, the just from the unjust, the victims from the perpetrators, the humans from the 'non-human') – then we are complicit in 'his' violence, because of its very humanity.[40] Hence, if gender works to keep the beast and violence separate from 'us', can we then degender[41] sexual violence and see it as human, and live with that?

These are the questions/difficulties/fears with which we are left struggling. The irony is that perhaps we fear being human more than we fear the 'bestial man' – a fear born out of our inherent implication in violence. It seems we fear the rapists less than we fear that we could become one, or that 'he' is I …

2 | 'Rape as a weapon of war'?

Wartime rape is 'a military tactic, serving as a combat tool to humiliate and demoralize individuals, to tear apart families, and to devastate communities'. (UN Action 2007)

In today's global political climate, to claim that wartime rape is a strategy or tactic of war is seemingly to state the obvious. In the previous chapter, we introduced and then critically analysed the available framings for understanding sexual violence and its subjects, including the Rape as a Weapon of War narrative. This narrative is embedded in and overlaps with the generalized 'Gendered' Story of sexual violence recounted there. In this chapter, we turn our gaze more specifically to the discourse of Rape as a Weapon of War in order to further understand its composition and, importantly, its purchase. How can rape be a weapon of war? What sorts of assumptions are needed to make this claim/explanation possible? And why is this framing of sexual violence so seductive and so prominent?

To recap: while other narratives of wartime rape surely resonate for many in various contexts, the storyline of Rape as a Weapon of War has become the most prevalent framing for understanding and redressing conflict-related sexual violence globally.[1] The widespread embracing of this narrative has undoubtedly been important for breaking with the view of rape as a tragic but inevitable outcome of war. Indeed, the Rape as a Weapon of War framework has also built upon, as well as made possible, novel and, in many cases, informed and pertinent accounts of wartime rape.[2] It has thus become the dominant explanatory framework within the research community,[3] the global policy community and the media. Rape as a Weapon of War can therefore be seen as the overriding general (and often universally applied) theory of wartime sexual violence, owing to its political and intellectual appeal and timeliness. Hence, taking the policy renditions, the media reporting and the academic (theoretical) explanations together, we can speak of it as a dominant discourse, which offers a grid of intelligibility for understanding and acting to prevent and remedy wartime sexual violence.

However, as we argued in the previous chapter, often what we see so clearly in the 'obvious' emerges in part through the concealment of a host of assumptions, logics and exclusions. Why speak of Rape as a Weapon of

War as a discourse? What can we gain by offering a critical discursive reading of the Rape as a Weapon of War narrative?[4] We suggest that by looking at the familiar terrains of policy, the media and academic writings,[5] and interrogating the underlying grammar that upholds the claim that rape is a weapon of war (against the backdrop of our analysis of sex-gender-violence), we can further comprehend why and how this framework is so appealing. Importantly, a critical reading allows us to reveal and interrogate the limits of this particular framing, despite its allure, and to begin to imagine a slightly different picture of rape and its subjects.

As a pre-established framework for describing wartime rape in all settings, the Rape as a Weapon of War narrative is frequently offered up as if it were somehow self-explanatory (particularly through its implied universalized storyline of gender as something we already know and understand). Currently (especially since the authorization of the UN Security Council's resolutions and commitment to combating sexual violence), evoking Rape as a Weapon of War as simultaneously a characterization of and an explanation for certain violent acts and consequences requires little argumentation or support in grounded empirical research. Sceptical and quizzical voices questioning its validity and generalizability, or even those calling for context-specific nuances, risk being framed as retrograde and as supporters of patriarchy who thwart the struggles to alleviate the plights of women in war. Hence, simply put, we join a growing number of critical scholars who feel the impetus to swear in the church of prevailing feminist politics, in order to call attention to its confines.

We therefore offer a critical reading of the discourse of Rape as a Weapon of War in order to make it visible and study its scaffolding against the backdrop of the overlapping 'Gendered' Story and the 'Sexed' Story that haunts it, recounted in the previous chapter. We do so in order to better understand the appeal of the narrative of Rape as a Weapon of War in the face of the violence of widespread and brutal conflict-related rape. This appeal, we suggest, resides in its inchoate promise that the bestial violent sex evoked in the 'Sexed' Story and (ironically) reproduced in the 'Gendered' Story can be hampered; criminals will come to justice, wartime rape can be eradicated, or at least largely prevented or avoided, and sexual violence can be controlled, managed and depoliticized. Before we further unpack the Rape as a Weapon of War discourse, however, we will briefly outline our method for analysis.

Reading the Rape as a Weapon of War discourse

Our aim here is to pick through and interrogate the main assumptions that underpin the weapon of war narrative which enable it to 'work' – that is, to make sense. One way of getting at these assumptions is by identifying the main components of the discourse, its privileged signs, 'nodal points', which glue the overall story together. (Think of an albeit floating bright star around which other stars in a constellation gather. The meaning of the constellation is dominated by the bright star's imparting the proper position and signification to the other stars in the group.[6]) This affixing process is not self-evident; how certain relations of signs are configured to constitute a constellation of seemingly definitive meaning has everything to do with the workings of politics and with political imaginaries.

We have found that the discourse of Rape as a Weapon of War revolves around four main interrelated points, which organize its narrative: 1) strategicness, 2) gender, 3) culpability and 4) avoidability.[7] Instead of giving equal weight to all four points, we choose 'strategicness' as the point of departure for our unravelling because we see it as key in the unfolding of a credible and cohesive storyline. In so doing, we are able to ask how strategicness is imbued with meaning, and how the notion of 'strategicness' actualized relates to a certain framing of gender, culpability and – importantly – avoidability. By querying the composition of strategicness, we are able to discern what may be excluded or expunged in order for sense to be rendered and seeming coherence established.

So where do we look for the Rape as a Weapon of War discourse? Discourses that reflect the 'state of knowledge' at particular times (and spaces) appear across a range of texts (Hall 1997b: 44). As noted above, we sought texts in the familiar terrains of policy, the media and academia. One of the tricks of conducting credible discourse analysis is, of course, choosing and delimiting the texts that are to represent the discourse which is the subject of analysis.[8] We have located the discourse of Rape as a Weapon of War in many different types of texts[9] (policy reports, academic literature, UN resolutions, media reporting).[10] Hence, we have sought to represent a 'discursive formation',[11] the main plot of which is regularly retold in global policy debates and media reporting, as well as in recent academic literature (Foucault 1972). Furthermore, its traces arguably constrain the storylines of (feminist) scholarship on rape in war more generally.[12] We draw on examples from all three of these terrains in our analysis, but delineate the differences when pertinent. When deconstructing the discourse through analysing particular texts, however, we primarily use examples of statements from the media, most notably the *Guardian* and the *New York Times*, and (global) policy documents.

In sum, the remainder of this chapter unpacks the discourse of Rape as a Weapon of War in light of our analysis of the 'Sexed'-'Gendered' Story, in which it is embedded. In order to better grasp how it is possible that rape is a weapon of war, we query some of the hidden assumptions, logics and, importantly, exclusions that impart sense to its storyline, and explore the (relations of) difference that inform its 'nodal points' and their associated patterns of meaning. Relatedly, we briefly discuss the subjects produced, as well as the conditions of possibility for these subjects. Because we have found that strategicness is so central in imparting sense to the discourse and to ensuring the promise and appeal of the ultimate avoidability of wartime sexual violence, we focus our analysis on the work that strategicness (as it is actualized in the discourse) does in making rape, as a weapon of war, possible and avoidable. As we shall see, the discourse of Rape as a Weapon of War clearly comprises a common main thread woven out of certain assumptions about gender (the 'gender' thread is also entwined with other threads emerging from the different nodal points we identify), and the relationship between sex and gender, more fully discussed in the previous chapter. Its very cogency and purchase are built on this 'Gendered' Story. Throughout our analysis, we therefore necessarily weave in a discussion of the relationships between strategicness, gender, culpability and avoidance in order to explore how they mutually inform each other and bind the storyline of Rape as a Weapon of War together.

Strategicness

Terrorism, torture, bombing – the most horrific acts are openly decried in times of war. Yet one particularly pervasive atrocity has been shrouded in a conspiracy of silence. US Secretary of State Hillary Clinton called it 'evil in its purest form' on her recent high profile Africa visit. This is the military tactic of mass rape. Absent from ceasefire agreements, not subject to disarmament programmes, and rarely mentioned at the peace-table, it is a war tactic that lingers long after the guns fall silent. Its legacy has shattered civilian lives and livelihoods, shredding the social fabric and devastating prospects for durable peace. In this decade alone, its toll has been staggering: claiming over 200,000 victims since conflict erupted in Eastern Congo. Yet you will not find these victims on official lists of the 'war wounded'. Unlike landmine or shrapnel injuries, its scars are invisible. In the arsenal of any armed group, this is the only weapon of mass destruction for which societies blame the victims, rather than the attackers. And though it is a war crime, it more often leads perpetrators to the corridors of power than to the cells of a prison. (Hilde F. Johnson, Co-chair of UN Action Against Sexual Violence in Conflict, 29 September 2009)

So don't think of wartime atrocities as some ineluctable Lord of the Flies reversion to life in a natural state but as a calculated military strategy. We can change those calculations by holding commanders accountable. (Kristof 2010b)

In the above quotations, rape is cast as a (strategic) weapon or tactic of war. These statements convey to us that wartime rape is intentional, following a certain rationality, and devised to effect particular outcomes. A report on sexual violence in Sierra Leone reiterates these generalized truths and explains: 'Rape as a Weapon of War serves a strategic function and acts as an integral tool for achieving military objectives' (Human Rights Watch 2003: 53).

While accounts of the strategic functions of rape differ in wordings and typologies between writers, and in light of the very different contexts that they address, they also overlap. While some refer to rape as a 'weapon' of war (or as a 'martial weapon'; Card 1996), some use the term 'strategy'. Others, more recently, have referred to it as a 'tactic' of war. Often, however, these terms are used interchangeably; little difference occurs in terms of the overall meanings imparted in the discursive formation (Howarth et al. 2000). As noted above, discourses are made up of numerous texts and genres. The privileged points in a discourse need not be – and often are not – identical in each text and genre.[13] What is at stake here is not a genealogy of the different meanings being imparted to these concepts in diverse fields, such as military theory or organizational theory.[14] The importance, for our purposes, has to do with asking how a certain idea of 'strategicness', whether it is conveyed through the term weapon, strategy, or tactic, is made possible and how a certain picture (Pin-Fat 2000) of strategicness is actualized in and helps make the discourse 'work' as a seemingly coherent story. Of course, differences between the terms are also meaningful, as we will also discuss below. However, the ways in which these terms are being used in the discourse has much more to do with their (political) purchase than with accurate correlation with their use in military settings.

What is meant by 'strategic'? In an interview in *The Nation* magazine, Margot Wallström, the former UN Special Representative of the Secretary-General on sexual violence in conflict, 'describes sexual abuse as a weapon of war, targeting not only women and girls but also men and boys, as planned and systematic, designed "to control the territory, to instil fear, to terrorize the population"' (quoted in Crossette 2010). In relation to the context of the DRC, the prevailing framework for understanding is that rape is used as a systematic and strategic weapon or tactic of war. For instance, the Special Rapporteur of the UN, Ertürk, concluded in 2008 that rape is 'used systematically in operations against civilian populations' (Ertürk 2008: 10) and is 'systematically employed to intimidate the local population' (ibid.: 8). Such framings are prevalent in Wallström's later description (2010b) of the widespread instances of sexual violence in the DRC:

The atrocities that are committed daily against women and children will leave a devastating imprint on the Congo for years to come. We have seen this elsewhere. In places where sexual violence has been used as a tactic of war, the consequences spill over into the peace. Where sexual violence has been a way of war it can destroy a way of life. Children accustomed to rape and violence can grow into adults who accept such behaviour as the norm. Rape is shattering traditions that anchor community values, disrupting their transmission to future generations. For the women of Walikale, peace is not a treaty, a resolution, or a conference but simply the peace of mind to live and work without fear. For these women justice delayed is more than justice denied – it is terror continued. (Ibid.)

In the policy world and the media, framing rape as 'strategic', a 'tactic', 'systematic' and 'planned', appears quite straightforward. Yet why and how wartime rape is considered 'strategic' in relation to the specific warscape in which sexual violence occurs remains often largely unexplained (why, for instance, does rape destroy a life; in what ways does it shatter traditions?). Instead, when explanations are given at all, they are generally rather formulaic and based loosely on the commonly accepted academic account as written through the 'Gendered' Story, in its various renditions. As we will further explore below, the 'strategicness' of wartime rape is assumed and implied through its relation to assumptions about gender relations, culpability and the possibility of avoidance.

In academic literature, explanatory frameworks are often more explicit, although they also depend upon the associations set in motion through chains of signification. Inger Skjelsbaek, for instance, outlines perceptions of the 'strategic effect' of sexual violence as a weapon of war:

a) it reaffirms militaristic masculinity; thereby focusing on the perpetrator, b) attacking the ethnic/religious/political identity that the woman is seen to embody, thereby turning the focus on to the victim, and c) masculinizing the perpetrators by empowering their identity and feminizing the victim by victimizing his/her identity, thereby focusing on the symbolic interaction between the perpetrator and the victim. (Skjelsbaek 2012: 89)

Most explanations of the 'strategicness' of wartime rape rely on a broad notion of strategy that does not necessarily entail that rape be a direct order for it to be strategic (Allen 1996; Card 1996; Gottschall 2004: 131; Littlewood 1997; Skjelsbaek 2001; Stiglmayer 1994; Samset n.d.). As well as direct orders, implicit condoning or encouragement of rape can also serve strategic purposes (Isikozlu and Millard 2010: 9; see also Lindsay 2005, cited in Isikozlu and Millard

47

2010: 9). Furthermore, individual rapists can serve strategic purposes even though they might not intend these purposes. Card, for instance, admits the complexity and multiple motives for individual acts, even though she relies on the notion of authority manifested in orders being sent along a chain of command.[15] In short, wartime rape is described (albeit in different ways) as a strategy of war as a 'martial weapon' (Card 1996), which enables armed groups to achieve military and political goals. How different authors understand these political purposes differs, of course – especially in light of the difference in the contexts of armed conflicts. Bernard, for instance, offers an overview of different generalized political purposes that sexual violence can serve:

> first, it facilitates ethnic cleansing by increasing the incentive to flee; second, it demoralizes the opponent; third, it signals an intention to break up society; fourth, it inflicts trauma and contributes to psychological damage by the opposing side; fifth, it gives psychological benefits to the perpetrators; and finally sixth, it inflicts a blow against the collective enemy by striking at a group with high symbolic value. (Bernard 1994: 35–9, cited in Skjelsbaek 2010: 37)

Some scholars argue that in order to further its aims, the military makes use of the notion of rape as a result of biological heterosexual urges (the 'Sexed' Story, as we saw in Chapter 1), thereby excusing and naturalizing violence against civilians as a regrettable, unintended effect. Seifert explains as follows:

> By using words like unforeseen or inadvertent, civilian victims are reduced to insignificance in the context of the conflict, and their suffering is disparaged [...] From an analytical point of view, such an approach obscures the fact that in reality the suffering of the civilian population which consists, as must be emphasized again, largely of women, constitutes a crucial element of war. (Seifert 1996: 38)

Others focus (also) on the ways in which patriarchal gender relations facilitate the effectiveness of a war waged on women's bodies for the purpose of furthering political, ideological or economic goals (see, for instance, Alison 2007; MacKinnon 1989; Niarchos 1995).

Undoubtedly, evidence of the widespread and strategic aspects of sexual violence in Rwanda (through, for example, the hate media Radio Television Libre des Milles Collines), and in Bosnia-Herzegovina in rape camps, has been well established both in international tribunals and in excellent academic and policy research.[16] However, in general terms, the empirical proof provided to support the argument of rape being strategic is often its widespread occurrence. A slippage in reasoning occurs here which extrapolates from what was arguably the case in Rwanda and Bosnia (the widespread aspect of sexual violence as part of ethnic cleansing) and a circular reasoning emerges. For example, a common argument is that sexual violence is simply so systematic and pervasive that because of its pervasiveness it must be part of a conscious policy (see Seifert 1996). In remarking on her overview of articles and publications addressing sexual violence in the 1990s, Skjelsbaek comments on the general consensus in scholarly literature: 'There is strong consensus that sexual violence is being used as a weapon of war [...] The use of sexual violence in the war-zone is simply too widespread, too frequent and, it seems, too calculated and effective for it not to be part of a larger political scheme and hence a weapon of war' (Skjelsbaek 2010: 30). This is echoed in policy reports, as well as the media, in, for example, the oft-repeated statement that the DRC is the 'rape capital of the world'.[17] The implied reasoning is that the occurrence of 'mass rapes' must mean that they are systematic and strategic. In the case of the DRC, evidence supporting this claim is, if anything, mainly anecdotal (see Chapter 3).[18]

We now turn to a brief discussion of the terminology of 'strategicness'. For the impatient reader, the following discussion may appear anomalous in the analytical mood of this chapter. It is included to provide an overview of how the notion of the overarching discourse on sexual violence in the DRC is being conveyed in different texts, in order to lay the groundwork for the subsequent analysis in the following sections.

Terminology: the language of strategicness What is the language used to convey strategicness? The term 'weapon of war' came into use during the trials in the Balkans and is still commonly used by popular media, UN agencies and academics alike to describe/characterize wartime rape.[19] Through the groundbreaking work at the ICU in 1993, it is well established that sexual violence in war constitutes a war crime (Henry 2011). This has lent credence to the notion that conflict-related sexual violence is indeed a 'weapon of war'. This notion now enjoys a well-established familiarity in common characterizations of sexual violence in armed conflicts. By 1993 Amnesty International,[20] as well as other NGOs, including media outlets like ABC and the *New York Times*, were already referring to rape as a 'weapon of war', albeit without the certainty that seems to have increasingly characterized reporting in the subsequent decade (Amnesty International 1993; EC Investigative Mission 1993; New York Times 1993). Nicholas D. Kristof of the *New York Times*, who has been one of the most consistently vocal journalists covering the topic, explains as follows:

The world woke up to this phenomenon in 1993, after discovering that Serbian forces had set up a network of 'rape camps' in which women and girls, some as young as 12, were enslaved. Since then, we've seen similar patterns of systematic rape in many countries, and it has become clear that *mass rape is not just a by-product of war but also sometimes a deliberate weapon.* (Kristof 2008; emphasis added)

Indeed, in a study of newspaper articles from the *Guardian* and the *New York Times* (2003–11), we found that many articles covering wartime rape spoke of it (or quoted others speaking of it) as a weapon of war.[21] In a comprehensive review (conducted April–December 2011) of available policy documents dealing with conflict-related sexual violence, we also found that a vast majority of policy documents did the same.[22] Indeed, in policy documents, the strategic component of wartime rape that renders it a specific type of 'weapon'[23] that demands international attention and retribution figures centrally.[24]

Furthermore, there were a few scholarly articles referring to Rape as a Weapon of War written as early as 1993, and the numbers have increased substantially since then.[25] In sum, the characterization of Rape as a Weapon of War has been commonly (and increasingly) used by popular media,[26] UN representatives and agencies, NGOs and academics.

The use of the term 'strategy of war', however, is more limited than that of 'weapon of war'. At the time of writing (2012), it is not easily found in official UN documents (unless one includes the use of 'strategic tactic'; the term 'tactic' will be discussed below). Occasionally, the term 'strategy of war' will appear in newspaper articles, with reference to a UN source, although this source is often not clear[27] (Kristof 2008; Simons 2009, 2010). According to our study of the media mentioned above, 'Strategy of war' is also sometimes used in reporting, although 'weapon of war' is much more common.[28] It is most prominently employed in connection with the global media and policy attention to the wars in the DRC, and with a backwards glance predominantly at the conflicts in the Balkans and Rwanda (EC Investigative Mission 1993; Human Rights Watch/Africa and Human Rights Watch Women's Rights Project 1996; Isikozlu and Millard 2010). Rape as a 'strategy of war' has not often been used by the UN officially (but has appeared in others' work that quotes UN officials, namely Stephen Lewis, a former UN envoy for AIDS in Africa[29]).

Academic usage of 'strategy of war' dates back to at least 1993 (see Swiss and Giller 1993) and is frequently used in discussions of conflict-related rape. Some scholars have differentiated between the concepts of 'weapon' and 'strategy' (Farwell 2004: 393). Koo, for instance, defines war rape as a weapon because it 'attacks women's physical and emotional sense of security while simultaneously launching an assault, through women's bodies [...]' (Koo 2002: 525).[30] As a strategy, rape is portrayed as a sanctioned, systematic means of attaining specific political objectives. The preference for the notion of weapon, rather than strategy, could be that it better connotes a sense of mastery and management over violence and the avoidance of violence, which will be further discussed below. Furthermore, the notion of weapon allows for the call for regulation of rape *as* a weapon, piggybacking on draft regulations against other destructive weapons used in warfare, such as biological weapons or chemical weapons.[31] We can see this reference in the following statements:

> There is an urgent need to disseminate the fact that sexual violence – whether a single act or concerted campaign – is categorically prohibited under international law. For communities, mass rape is a weapon of mass destruction like any other. It ranks among the grave breaches of international humanitarian law, reflected in the 1998 *Rome Statute of the International Criminal Court*; 1949 Geneva Conventions; and jurisprudence of the international criminal tribunals for the former Yugoslavia and Rwanda. Today, sexual violence is an international crime – not the timeless 'collateral damage' of war. (UNDP 2008: 2)

> World leaders fight terrorism all the time, with summit meetings and soundbites and security initiatives. But they have studiously ignored one of the most common and brutal varieties of terrorism in the world today. This is a kind of terrorism that disproportionately targets children. It involves not W.M.D. but simply AK-47s, machetes and pointed sticks. It is mass rape – and it will be elevated, belatedly, to a spot on the international agenda this week. (Kristof 2008)

However, the adoption of the language of rape as a 'weapon' has also resulted in numerous jokes among those working with sexual violence in the policy community who ridicule sexual violence activists. A common line of humour is as follows: 'so if it is a weapon, how should it be disarmed, then?'[32] The prevalence of these kinds of jokes, even at high levels, was supposedly one of the reasons why UN Action in 2010 declared that they were now using the term 'tactic' in preference to 'weapon'.[33]

Related to the term strategy, the notion of tactic also often figures in policy texts, perhaps most prominently the Security Council Resolutions 1820 and 1888.[34] Additionally, most other policy documents, with minor variations in wording, refer to UN Action's statement, under the heading 'Sexual violence as a weapon of war', that wartime rape is 'a military *tactic*, serving as a combat tool to humiliate and demoralize individuals, to tear apart families, and to devastate communities' (UN Action 2007: 5). For instance, we can see how this notion of tactic appears in the following statement, thus reiterating what has rapidly become an accepted truth: 'women and girls are particularly targeted by the use of sexual violence, including as a tactic of war to humiliate, dominate, instil fear in, disperse and/or forcibly relocate civilian members of a community or ethnic group' (Security Council Resolution 1820, quoted in UNIFEM, Stop Rape Now and United Nations 2010: 5).

The use of *tactic* also appears in the media, as a shorthand (often with no further explanation).[35] Furthermore, 'strategy' and 'tactic' can be seen as intimately interrelated and interdependent. Whereas strategy connotes a wider programme or plan designed to implement a (political) goal of warring – in this case often 'humiliation of the enemy' – tactic can be seen as the technique (rape) of implementing the strategy: 'humiliation of the enemy'.[36]

However, these distinctions are often quite muddied, and importantly, the sense conveyed is that rape is intentional and for a particular (strategic) and political purpose of warring, be it through use of the term weapon, strategy or tactic, or other variations, such as the frequently used notion of rape being 'systematic' (Wallström, quoted in Crossette 2010). Hence, while the categorizations and wording differ,[37] together these terms convey and actualize 'strategicness', as functionality and intent.

Unpacking Rape as a Weapon of War

Now that we have explored the terminology used to connote strategicness, we turn to the central question of meaning, and the moves through which meaning is made possible.

The rational strategic actor How is Rape as a Weapon of War made possible through actualizations of strategicness and the attending notions of culpability, gender and avoidability? If we recall Kristof's appeal in the opening quotes of this section, we can see how he cautions us *not* to think of rape as some 'ineluctable Lord of the Flies reversion to life in a natural state, but as a calculated military strategy' (Kristof 2008). The message seems to be clear and simple: rape is a weapon that is wielded for particular purposes; it is a weapon of choice wielded by rational modern subjects, be they military commanders or soldiers on the ground. These rational subjects intend evil for particular purposes, and therewith can be held accountable.[38] However, upon reflection, we wonder why Kristof felt compelled (as indeed he should) to guide us away from our supposed initial interpretative impulses. What might his forewarning tell us about how strategicness is constructed (and what is necessarily expunged), and what the notion of strategicness does in the discourse of Rape as a Weapon of War?

If we revisit two statements below (which are indicative of the larger discursive formation) and look for the ways in which associations and inclusions/ exclusions are at play, as well as explore what types of subjects are made possible, we can begin to understand why we are asked *not* to think about a fictitious 'natural state'. We find the following formulations:

> Sexual Violence as a Weapon of War: Conflict creates the climate for rampant sexual violence. Sexual violence has been dismissed as random acts of individual soldiers. But in armed conflict, rape is also often a military tactic, serving as a combat tool to humiliate and demoralize individuals, to tear apart families, and to devastate communities. Armed forces use sexual violence as the spoils of war for soldiers who see the rape of women as their entitlement. Lawlessness allows perpetrators to act with impunity and leaves survivors with little to no recourse. (UN Action 2007: 5)

> Sexual violence in conflict is a serious, present-day emergency affecting millions of people, primarily women and girls. It is frequently a *conscious strategy* employed on a large scale by armed groups to humiliate opponents and destroy individuals, as well as whole societies. Sexual violence during conflict remains vastly under-addressed due to weak national protection mechanisms, inadequate judicial redress and piecemeal services for survivors. Many still view sexual violence as an inevitable, if regrettable, consequence of conflict and displacement – an attitude which encourages impunity for perpetrators and silences survivors. Yet rape during conflict is a war crime, crime against humanity, act of genocide and form of torture. (UN Action 2010)

In a similar way to that we saw in the previous chapter in our analysis of the generalized 'Gendered' Story, these statements make sense in part through their contradistinction to the previously dominant, and still existing (but increasingly marginalized), 'Sexed' Story. Importantly, instead of being unintended, rape is framed as intended – for particular purposes: 'a conscious strategy'; 'a military tactic, serving as a combat tool to humiliate and demoralize individuals, to tear apart families, and to devastate communities'; 'to humiliate opponents and destroy individuals, as well as whole societies'. According to this storyline, certain action derives from certain intention and clear ethical distinctions can be rendered (and responsibility borne) (Schott 2004, 2011).

Instead of being driven by 'natural', even animalistic, sexual urge, rape in war is about pursuing military and political goals. Biological drives (for the most part) are explicitly evacuated from this reasoning: 'random' and 'inevitable, if regrettable' acts (of sexuality/evil run amok?) destined by biology and let loose in the climate of warring are replaced in this discourse by conscious and purposive tactics and strategies. If we recall our discussion in Chapter 1 of the distinctions drawn between humans and beasts in the 'Sexed' Story, we can again intimate the desire to place rapists in the realm of the rational human who is responsible for his actions. Conscious and purposive acts, as we shall further discuss below, enable the establishment of guilt (as rape is a 'war crime, crime against humanity') and, ultimately, the prevention of such a 'strategy'.

The primacy of biology, however, haunts these renditions of the Rape as a Weapon of War both through its exclusion and evocations in words like 'yet': 'Sexual violence has been *dismissed* as random acts of individual soldiers. *Yet* in armed conflict [...]'. Additionally, its subplots even linger in the main story. In the first quotation (from UN Action), for instance, the second half of the quote seems to contradict the first part. 'Armed forces use sexual violence as the spoils of war for soldiers who see the rape of women as their entitlement.' Here the logic, which renders comprehensible 'war booty' or the 'spoils of war' as an explanation for wartime rape (rape results from men's heterosexual desires that get deferred and then fulfilled as a reward after battle), troubles the plot of the Rape as a Weapon of War narrative. Nonetheless, the heavy work done by the terms 'tactic' and 'combat tool' in this story holds sway, and the primary meaning imparted is that rape is a strategic and lethal weapon, which is wielded consciously by perpetrators in control of both their bodies and the plight of their victims.

What kinds of subjects are assumed (and produced) through such texts? Clearly (as we discussed in the previous chapter), two main characters dominate the plot: the women/victims and the soldiers/rapists/perpetrators. We flesh this out further through reading examples of the discourse of Rape as a Weapon of War from Human Rights Watch reports, the first addressing the genocide in Rwanda, and the second the war in eastern Congo:

> The humiliation, pain and terror inflicted by the rapist is meant to degrade not just the individual woman but also to strip the humanity from the larger group of which she is a part. The rape of one person is translated into an assault upon the community through the emphasis placed in every culture on women's sexual virtue: the shame of the rape humiliates the family and all those associated with the survivor. Combatants who rape in war often explicitly link their acts of sexual violence to this broader social degradation. (Human Rights Watch/Africa and Human Rights Watch Women's Rights Project 1996)

> Soldiers and combatants raped and otherwise abused women and girls

as part of their effort to win and maintain control over civilians and the territory they inhabited. They attacked women and girls as representatives of their communities, intending through their injury and humiliation to terrorize the women themselves and many others. (Human Rights Watch 2002: 23)

The 'victim' in the first HRW text on Rwanda is a woman/girl who is defined through her gendering as female – whose 'sexual virtuousness' represents her community as a whole. Her 'shame' as a result of rape infects the family and all with whom she is associated, and ultimately the community as a whole, as she acts as a symbol of her culture. We can see how the combatant who rapes 'explicitly' links his act to 'broader social degradation'. In the subsequent quotation from the DRC report, the perpetrator 'intends' through his injury to terrorize the 'women themselves' and many others, through the powerful workings of assumed gender discourses.

The notions that these goals could possibly be achieved through the strategy of rape depends upon gendered assumptions about how men and women are (and are not), heterosexuality is (and is not), and how gender works (and does not work) in and upon communities and individuals. The goals 'achieved' by rape are further described: rapists 'attacked women and girls as representatives of their communities, intending through their injury and humiliation to terrorize the women themselves and many others'. This statement convincingly lets us know that a whole host of 'injuries' and 'humiliations' (which we can imagine through listening to the implied harms through our gender-tuned frequencies) are set in motion through the terror that women, *as* representatives of their communities, are subject to. We know, generally speaking, through the framing of the 'Gendered' Story, who and how such women are, and what the significance of acting as a representative of her community means to her, her community and those who aim to destroy it through/with/and her.

Who is the perpetrator/'rapist' in these accounts? He is undoubtedly a heterosexual male who commits the torture/crime/act of rape for the specific purpose of 'tearing apart communities', etc., for strategic reasons. He emerges as a rational person in control of his actions, who is either following orders of military hierarchy or acting of his own volition (as it is unclear who, actually, is doing the ultimate 'intending').[39] Indeed, certain assumptions about military hierarchy as being ordered with clear chains of command and the successful production of 'docile' yet raping bodies (Foucault 1991) can be seen as a condition of possibility for this militarized subject; this will be further developed in the next chapter.[40]

While his acts may be brutal and the consequences for those he harms and what they represent dire, he nonetheless is doing it for a *reason* which lies beyond the unsavoury mixture of the need for sexual satisfaction and barbaric evil.[41] The picture of humanity portrayed in *Lord of the Flies* (Golding 1954) is thus held at bay and the (bestial, violent) subject that the rational subject *shall* overcome (Pin-Fat forthcoming) is seemingly left behind. Yet this uncontrolled and unreasoned animal lurks behind the rational subject, who is in control, can be punished for his crimes, and dissuaded from violence. Indeed, this lurking shadow intimates that the gendered subject who is taught/forced/governed to rape for strategic purposes is as violently out of (social) order as the sexed one, who is at once at one with, and controlled by, violence. The insistence that sexual violence is strategic, systematic, rational lulls us into thinking that the gendered violent subject resides in a moral world whose contours we recognize and as such can indeed be known, in control, punished, and reformed.

The culpable, punishable subject The notion of guilt and culpability, institutionally established through the International Criminal Tribunal for the former Yugoslavia (ICTY) in 1993 and the International Criminal Tribunal for Rwanda (ICTR)[42] in 1994, is crucial for rendering this narrative comprehensible. If the rapists or those who govern them are indeed in control and *intending* to harm for particular purposes, then they can be held accountable.[43] The notion that rape is not a crime ultimately produced by nature, but instead by gendered, social, power relations, produces a different picture of the burden of responsibility and accountability from that produced by different versions of the 'Sexed' Story. Rape moves from being a terrible derivative of nature, let loose through war, to being a war crime enacted for purposes beyond the specific act. The rapist who commits a war crime acts within the realms of the social, guided (or misguided) by allegiances and ethical compasses that hold him accountable as a rational, responsible subject who has, and makes, choices, even destructive poor ones – and can be punished for them.[44] The rapist who is ruled by the needs of his body acts not out of some injudicious view of strategic gains *within* a social order, but out of desire and 'natural' urges and performs somehow *outside* of society.[45] He acts without rationality (if we accept the rational as defined in opposition to the animalistic or the emotional, as discussed in Chapter 1) and therewith without ethics to (mis)guide him. Following this reasoning, he cannot be fully accountable within the realms of the social, and even the fully human.[46]

As we can see in the example below from an IRIN and OCHA report (Horwood et al. 2007), the act of rape shifts in the discourse of Rape as a Weapon of War from being an albeit terrible personalized, private, sexual act to a war crime. Indeed, this very shift is an important component in the purchase of this discourse:

> Despite its pervasiveness, rape is often a hidden element of war. Because the use is largely gender-specific and committed by men against women, it is usually narrowly portrayed as being sexual or personal in nature, as a private crime or as a sexual act. Rape, however, is sometimes part of a premeditated political or military strategy. Ignoring the fact that sexual violence against women and girls is used as a combat tactic trivializes what in reality is a war crime. (Ibid.: 37–8)

Guilt and the responsibility of the perpetrator, who is often constructed as a 'helpless victim of innate and ineradicable impulses' (Gottschall 2004: 135), surely invited punishment in the framing offered by the 'Sexed' Story. The Rape as a Weapon of War narrative, however, allows for an even more stringent and significant sense of responsibility, culpability and guilt, as war rape is now framed as a crime against humanity (to which the perpetrator belongs), a war crime, and regulated in international law.[47] Furthermore, guilt is not only reserved for those soldiers who have enacted violence, but follows the chain of command, so that those '*intending* through their injury and humiliation to terrorize the women themselves and many others' by governing (or not) sexual violence are held accountable.[48]

By rendering wartime rape a war crime, which shall be punished, the discourse of Rape as a Weapon of War enables a call for retributive justice and the end to impunity as both punishment and as deterrent. By focusing on rectifying impunity, 'warnings are served to perpetrators' and the victims of sexual violence are afforded a 'glimmer of hope' (see statement below). Margot Wallström explained as follows:

> I named Mayele specifically in my briefing to the Security Council several weeks ago, and his arrest sends a loud and clear message that impunity for crimes of sexual violence will not be tolerated. [...] We must seize the momentum of these arrests to begin turning the tide of impunity. The arrests must serve as a warning to perpetrators of sexual violence everywhere. And, we cannot underestimate the importance of such action for the victims and their communities. This represents a glimmer of hope for them. A moment of solace that the world is not blind to their plight. A possibility that those who brutalize them will ultimately be held to account. (Wallström 2010b)

In addition to the hope of being heard, and justice being served, the 'glimmer of hope' refers, it is intimated, to future potential perpetrators heeding these 'warnings' and choosing to act differently.

Additionally, the following statement by UN Action offers a clear example of the way in which this shift in framing is represented. It thus sets the stage for political and legal efforts to end the impunity, the silencing of the terror of rape:

> Mass impunity has kept rape off the historical record and under the security radar. Women and girls have been treated as second-class victims of a second-class crime. Sexual violence has accordingly been side-lined by the world's most powerful security stakeholders as the private, inevitable or opportunistic excesses of a few renegade soldiers. This myth plays directly into the hands of those who wield rape as a weapon of choice, being low-cost, high-impact and less scrutinized than murder or mass graves. For these girls, who live in the midst of their tormentors, at the epicenter of a rape crisis, justice delayed is more than justice denied – it is terror continued. (Johnson 2009)

In this statement, the myth that casts rape as 'private, inevitable or opportunistic excesses of a few renegade soldiers' has been debunked; instead, excesses emerge instead as chosen strategy (chosen because, in part, of the power of this myth). Terror will continue unless justice is served to those who 'wield rape as a weapon of choice'.

The emphasis on weaponry in some texts allows us to glimpse the move to regain mastery over the 'tools' of violence. Surely, the shadowy, bestial, violent subject does not master his 'tool' of violence – his penis, as well as his bloodlust – and is in fact subject to its power and desire. The rational (albeit violently gendered) subject who takes his place in the discourse of Rape as a Weapon of War must master his weapon if he is to be held accountable for his crimes and learn to refrain from using his tools destructively. We can glimpse the mastery implied in the following statement:

> Mukwege[49] says sexual assault is comparable to biological warfare as an extermination tactic. He says there is a policy to make fathers and children watch the rapes. To render the woman sterile, the rapists complete the brutality by firing a bullet into the vagina or shredding its walls using a rifle butt or tree branch. (Smith 2010)

In this statement, those who wield the weapons of 'biological warfare' – be they a penis (alluded to but not stated), a rifle or a 'tree branch' – control their weapons, and not vice versa. Through this control they are separated from the violence they commit, and are not, like their shadows, produced through and subject to violence. The rational subject, while unequivocally enacting horrendous violence, is thus seemingly separated from violence, and can perhaps learn to act otherwise. (And, as we recall from Chapter 1, we, as those who are to know him, can rest easily in that his shadows do not also lurk inside us.)

Avoidability and the promise of deliverance

> Still, we in the West too often find it easier to perceive rape as an accepted part of an unfamiliar culture rather than as a tool of war that we could help banish. Too often, the enemy becomes all Congolese men rather than men with guns terrorizing the Congolese people. By casting the chaos and violence as 'men vs. women' or dismissing the crisis as 'cultural,' we do a profound injustice to Congolese men. Rather than help, we send an implicit insult: It's a pity, but, well [...] it's just who you people are. (Shannon 2010)

'Rape is not an inevitable consequence of war', most policy texts assure us.[50] These words usually figure as an introduction to a description of horrible rape scenes, or testimonies witnessing horrendous violence and the havoc it wreaks on people's lives, livelihoods and futures. We read these words and are comforted. Our comfort may be uneasy, but it is nonetheless a necessary counterweight to, and condition for, our engagement in stopping sexual violence in war. The reality may be horrific, we tell ourselves, but something can be done!

The appeal of the discourse of Rape as a Weapon of War rests, we suggest, largely on its promises for change. It is often formulated in negative terms, as 'rape is not unavoidable' (for example, Johnson 2009, UN Action 2007 refer to rape and sexual violence as 'not inevitable'). Indeed, its purchase and institutionalization in the global policy community, best exemplified in the mandate of the UN Special Representative for combating sexual violence, rely on the framing of rape in war as avoidable and as an abhorrent condition that can be treated. 'A glimmer of hope', as Wallström pointed out to us, has shone through after years of silence and portrayals of rape as inevitable. The framework of Rape as a Weapon of War promises a brighter future for sexually abused women (and more rarely men) in conflicts not only because their plight will be heard and their attackers punished, but because future rapes can be heeded. Its appeal is therefore highly seductive.

We have already touched on how this notion of avoidability has been premised by assumptions about strategicness, gender and culpability, and constructed in opposition to the characterization of rape offered by the 'Sexed' Story (which offers little, if any, prospects for change). As Gottschall (2004: 135) argues, 'allowing biology a beachhead in the explanation of mass wartime rape seems, to those passionately committed to seeking solutions, like the first step towards surrender to inevitability'. How might we further understand the conditions of possibility for the promise of deliverance offered by the discourse of Rape as a Weapon of War?

The story of sexual violence and its subjects written in this discourse is a teleological one; sense is imparted through the organization of the narrative into a given past, present and future, with a beginning, middle and end, and clear, coherent, stable subjects who progress along its temporal trajectory (see Disch 2003: 264; Stern 2006: 192). For it to be credible, the promise of deliverance from sexual violence depends on knowing who its subjects are (as we have discussed above) and how the trajectory of the plot moves through time and arrives at its teleological endpoint. Let us consider the following quotation:

> Mass rape has thus graduated from an 'inevitable by-product of war' to a foreign policy priority. SCR 1820 demands the 'immediate and complete cessation by all parties to armed conflict of all acts of sexual violence against civilians'. This responds to the reality that sexual violence has acquired a strategic twist as a tactic of choice for armed groups. (UNDP 2008: 2)

In this statement, mass rape has 'graduated' from its previous status as an 'inevitable by-product of war'; it is not a foreign policy priority. Mass rape has grown up and entered the big league of high security threat. Or perhaps it is we who have developed and evolved in our understanding of mass rape; it is we who have grown out of sex and into gender. In either case, a story of certain liberal, progressive evolution is portrayed, which allows for the alleviation of mass rape to be imminent. In the following quotation, we see again how the temporal structure of the storyline is revealed through wording such as 'still':

> Many *still* view sexual violence as an inevitable, if regrettable, consequence of conflict and displacement – an attitude which encourages impunity for perpetrators and silences survivors. Yet rape during conflict is a war crime, crime against humanity, act of genocide and form of torture. (UN Action 2010; emphasis added)

'Still' refers back in time to a previously credible truth, which has been abandoned to a present-day certainty: 'yet, rape during conflict is a war crime'.

The political urgency of portraying sexual violence as strategic, a crime and ultimately avoidable as a seemingly natural progression is revealed in the following citation: 'But we cannot wait for peace to bring peace to the lives of women. We must insist that where sexual violence is planned and orchestrated as a tactic of war, it must be viewed as preventable' (Wallström 2010b).

The word 'must' is important in flagging how framing rape as avoidable and preventable is vital to the promise of 'peace to the lives of women'. We must *insist* that it be viewed as preventable. In so insisting, the discourse of Rape as a Weapon of War is rendered a hopeful framework upon which to foster engagement and build policy: to enforce liberal laws and reproduce, rein in, punish and reform liberal sovereign subjects who can be held accountable and eventually successfully responsibilized.[51]

The possible policy implications of the 'Sexed' Story (which we have seemingly left behind) appear, needless to say, not only as quite limited, but problematic and incompatible with the current official ethics and norms of warfare (see Chapter 3). While the 'Sexed' Story does not necessarily imply the lack of acceptable policy recommendations, such as increased disciplinary measures to curtail ('natural' heterosexual male) behaviour, many of those policies that have been put in place are nonetheless associated with providing a 'natural' non-violent outlet for the fulfilment of male sexual desires (through increased accesses to prostitutes, wives or girlfriends in deployment areas). (See, for example, Enloe 2000; Gottschall 2004.)

By contrast, the Rape as a Weapon of War discourse promises change and something new, which meets the urgency that our newly acquired awareness of rape as 'planned and orchestrated as a tactic of war' entails. New programmes, drawing on international law, designed to address impunity and to discipline unruly (yet nonetheless rational) subjects, are presented as both effective and ethically attractive. In line with the global discourse on fomenting 'good governance' (e.g. Fukuyama 2004; Rotberg 2003), mechanisms for promoting self-discipline are put in place in peace-building/reconstruction efforts and SSR in order to prompt locally responsible governance, including the end to impunity (Eriksson Baaz and Stern 2013b). UN Action's comments on the 'List of Shame', which is a 'powerful incentive for behavioural change', provide a good example of the alluring promise of redemption:

> A powerful illustration of this renewed resolve is the Council's recent decision to expand the so-called 'list of shame' on violations against children in armed conflict to encompass perpetrators of 'rape and other grave sexual abuse'. The list currently names and shames groups that recruit child soldiers. Being 'de-listed' has proven a powerful incentive for behavioural change. It should have a similar effect on those who commit sexual abuse, providing information that can be acted upon by Sanctions Committees or international courts. Exposing persistent violators puts the burden of sexual violence back where it belongs – with the perpetrators. (Johnson 2009)

Additionally, if rape is a weapon or a strategy its use, similarly to other weapons or strategies, can be regulated and controlled. This is especially so when seen in the context of the international engagement in redressing the widespread rape in the 'rape capital of the world'. Violence committed by the use of these 'weapons', for example, can be mitigated through the imparting of knowledge about human rights and gender training in SSR efforts. Quite simply, we can teach soldiers a less violent masculinity (see Eriksson Baaz and Stern 2010). Violence and the weapons that wield it are thus comfortably moved back into the control of security governance and neutralized through the responsibilization of the state (security and justice systems) via international security interventions (Abrahamsen 2004; Burchell 1996; Eriksson Baaz and Stern 2010, 2013; Hansson 2012; O'Malley 1996).

The governing of gender in society at large also performs in the discourse as a promissory note for the prevention of sexual violence. The logic is as follows. If we understand rape as a gendered weapon of war – as avoidable and not inevitable – it can be changed. To foster desirable change, we can make use of the considerable research and best practices in promoting gender equality; we can combat sexual violence through changing gender relations in society at large, and within the security sector specifically.[52] Efforts to improve gender equality therefore emerge as highly politically important. Consequently, the effects of patriarchy or unequal gender relations often figure as 'causes' that can be remedied. Indeed, the question of the difference between rape in 'peacetime' and rape in the midst of and integral to warring remains one that poses difficulties for scholars and policy-makers alike (see, e.g., MacKinnon 1989; Alison 2007; Card 1996). Although writers do indeed differ in their perspectives on the role that peacetime gender inequalities play in contexts of warring, there seems to be a general sliding – or at least an ambiguous relationship – between rape as product of warring and militarized gendered subjects, and rape as indicative of and produced in unequal gender relations in society in the discursive formation more widely.[53] The common rhetoric, nonetheless, that rape in war is 'a war against women'[54] allows space for the call for efforts to increase gender equality in society at large to prevent conflict-related sexual violence.[55]

Concluding thoughts

First, an important qualification: we want to emphasize that this chapter should not be read as a rejection of the discourse of Rape as a Weapon of War. Not only has this explanatory framework had a central role in rendering sexual violence a global security issue, it also offers a useful basis for understanding and redressing the widespread use of sexual violence in many conflicts. The rape camps in the Balkans (Human Rights Watch 2000) as well as the genocide in Rwanda (Human Rights Watch/Africa and Human Rights Watch Women's Rights Project 1996) seem to be concrete manifestations of the ways in which rape is used as a strategy to further political and military aims. However, as we will further address in the next chapter, conflict settings and armed groups differ in their intricate complexities. A universal and universalizing storyline can only fail in its remit to provide *the* answers and remedies. At best, it can provide some answers and prompt further questions – if room is allowed for reflection, research and inquiry.

In this chapter, we have therefore sought to interrogate the discourse of Rape as a Weapon of War, to examine its scaffolding and to ask what makes it work as well as why it is so appealing. In so doing, we have concentrated our analysis on how the actualization of strategicness, culpability, gender and avoidability renders a seemingly cohesive narrative, which holds much politico-ethical purchase in the efforts to understand and combat conflict-related sexual violence globally. Our critical reading has revealed, however, that that which 'is already known' is not as securely established as implied. Other excluded explanatory framings (such as the different renditions of the 'Sexed' Story) haunt its cohesion. Fissures in its own inherent logic reveal the shakiness of seemingly sure foundations. Despite its teleological seduction and political importance, the discourse of Rape as a Weapon of War remains unstable and its promise of deliverance precarious.

The Rape as a Weapon of War framework, while seductive, must surely be further interrogated, not only in terms of its own logics and the assumptions upon which these logics rest, but also in terms of the particular contexts in which conflict-related rape occurs. This is the subject of the next chapter.

3 | The messiness and uncertainty of warring

In the previous chapter we analysed the discourse of Rape as a Weapon of War, through critically exploring how strategicness and the attendant notions of culpability, gender and avoidability together produce a seemingly cohesive, credible and politically appealing narrative. In this chapter we revisit the notion of strategicness, except this time from a different angle. Drawing upon insights garnered from the sociology of violence and the military and research in the Democratic Republic of the Congo (DRC), as well as research material from other conflict/post-conflict settings, we further query the notion of rape as (somehow) inherently strategic in warring. We do so by highlighting some aspects of military organizations and warring that tend to be rendered invisible in the generalized story of the strategicness of rape. To be clear, we do not attempt to provide a full explanation of rape in war, but to point to gaps in the Rape as a Weapon of War discourse through highlighting other viable explanatory factors which this discourse neglects, linked to the set-up of military structures and the nature of warring.

We address three aspects in particular. First, we attend to the discursive nature of 'strategy', and demonstrate the ways in which notions of military strategicness, including the strategicness of sexual violence, vary depending on military contexts. Second, we turn to the workings of military institutions and highlight that, in contrast to the ascendant representation in the Weapon of War discourse, military institutions seldom (if ever) embody the ideals of discipline, hierarchy and control to which they aspire. Rather than reflecting strategic action, sexual violence in war can also reflect the breakdown and fragility of military structures. Third, we demonstrate how the messy realities of warring trouble notions of rape in war as a strategic weapon by attending to the workings of cycles of violence and the micro-dynamics of warring. Throughout, we will illustrate these three aspects through references to the most recent warzone in which the discourse of Rape as a Weapon of War has been particularly dominant: the DRC.

An initial clarifying note: as concluded in the previous chapter, the sense conveyed in the discourse of Rape as a Weapon of War is that rape is intentional and useful/functional, be it through use of the terminology of 'weapon', 'strategy' or 'tactic' (or other variations, such as the frequently used notion of rape being 'systematic'). It is this general notion of the *intentionality* and *functionality* of rape which is problematized in this chapter. For this reason, we will not embark on defining, or differentiating between, the various concepts used (weapon, strategy, tactic, etc.), or on a further genealogy of their use in the Rape as a Weapon of War discourse, as we did in Chapter 2.[1] This chapter therefore takes its point of departure in, and interrogates, the general intentionality and functionality implied (whether on the level of strategy, tactics or weapon).

The references to the DRC are based on our (and others') research into the newly integrated Congolese armed forces.[2] The 'half-brewed' (Verweijen 2013b) Congolese army faces a range of challenges. As we contend, violence committed against civilian populations, including sexual violence, should partly be understood as manifestations of these challenges.

In this chapter we argue that the occurrence of sexual violence in warring contexts does not necessarily imply that sexual violence is construed as strategic, or is encouraged by military commanders.[3] As we briefly discussed in Chapter 2 and as Gottschall notes, 'While mass wartime rape can surely result in the damage discussed above [demoralized populaces or fractured families], it remains possible that the supporters of strategic rape theory may be confusing the consequences of wartime rape with the motives for it' (Gottschall 2004: 132). As we further contend through the remainder of this chapter, reducing rape in a warring context to a function simply of strategy, encouraged by military commanders, accords too much rationality and intentionality to wartime violence. Intentions and meanings cannot be inferred from outcomes in any context, particularly so in the messy realities of warring.

The discursive nature of military strategicness

One problematic aspect of the discourse of Rape as a Weapon of War is that it somehow assumes the existence of a uniform universal military strategy that is shared by all military/armed groups in all contexts, in which rape is construed as (somewhat) inherently and objectively strategic. It must, we argue, be recognized that strategy and tactics, as well as the consequences of particular violent actions, are imbued with meaning discursively. While the consequences on the bodies targeted are undoubtedly tangible and very 'real' indeed, the meanings attached to these violent actions and the interpretations of their origins, consequences and their potential strategicness will vary depending on discursive contexts. Hence, the results of a particular line of violent action might be construed as strategic by one actor and non-strategic by another.

Moreover, as argued by Clausewitz (1982 [1832]; Strachan 2007), it has to be remembered that a successful military strategy may be a means to an end, but it is not an end in itself. Hence, military success and a temporary defeat of the enemy do not necessarily imply that the political objective behind the military operation has been fulfilled. The crucial issue is the implications of this (temporary defeat) and whether these contribute to the fulfilment of the political aim of the military intervention. There are ample examples where military success/defeat over the enemy has not brought a realization of main objectives, but instead has generated new resistance and strengthened insurgency activity. This often results in a retrospective rewriting of the action, where it is reinscribed as non-strategic instead of as strategic. Hence, there is no objective definition of what constitutes a strategic military action that is *outside* competing discourses. Writing rape in war as inherently strategic obscures the ways in which military discourses of strategicness vary from one military actor to another, and from one context of warring to another. Let us below provide an example from counter-insurgency strategies – COIN, i.e. actions taken by a government to control or suppress a rebellion (insurgency).

Shifting meanings of COIN and the strategicness of rape Counter-insurgency is generally considered one of the most difficult forms of military intervention because of the tremendously intricate, but crucial, task of differentiating between insurgents, unarmed supporters and civilians who do not support the insurgency. Hence, counter-insurgency operations entail great risks that 'innocent civilians' may become victims. The current dominant and officially accepted strategy emphasizes the importance (strategicness) of being able to differentiate between insurgents and non-insurgents. This depends on creating good relations with the population through compelling them to cease supporting the enemy, and instead to support the counter-insurgent with information and other resources. This approach is captured in the expression 'winning the hearts and minds' of the population (cf. Aylwin-Foster 2005; Dickinson 2009; Nagl 2005).[4]

However, not only does history provide ample examples of the utter failure of this strategy in practice (a failure which also partly reflects the fact that the slogan is used as war propaganda, rather than reflecting the 'real' military strategy of a military operation), the universal effectiveness of this strategy is also questioned. For example, the military strategist Van Creveld (2008: 268) has argued that 'the first, and absolutely indispensable, thing to do is throw overboard 99 per cent of the literature on counterinsurgency', 'since most of it was written by the losing side, it is of little value' (ibid.); he emphasizes the disadvantages of a strong counter-insurgent, describing the counter-insurgency/ insurgency relationship through an analogy of a fight between an adult and a child; the insurgent/child risks receiving all the support and sympathy and the adult/counter-insurgent, all the blame. In order for the counter-insurgent to succeed, it has to be equipped with impeccable intelligence services and highly disciplined soldiers who are able to exercise restraint even when exposed to severe provocation. Unless this (unattainable?) capacity is in place, (extra) coercive strategies can be considered more effective, according to Van Creveld. He uses the example of the Syrian government's strategy in the Hama massacre in 1982, in which between ten and twenty-five thousand people were killed.[5] This strategy, which Van Creveld summarizes in five rules, including the necessity of cruelty in some situations, the importance of decisive action early on and the advantages of killing too many rather than not enough in one single strike (since repeated strikes can be detrimental to troop morale; ibid.: 241–5), stands in stark contrast to the strategy of 'winning the hearts and minds of people'. However, history provides ample examples of political and military actors – other than the al-Assad regime – that construe this type of extremely coercive counter-insurgency as effective, hence strategic. In sum, military discourses that define what is strategic and what is not vary from one political and military actor to another and from one context of conflict to another.

This also applies to the question of the strategicness of wartime rape. Studies of armed groups and military units demonstrate that commanders (or political leaders) frequently perceive rape as counterproductive and therefore try to minimize, rather than encourage, the rape of women by their troops (see Goldstein 2001; Gottschall 2004; Wood 2009, 2010). One of the most widely cited examples of this is the famous 'rape of Nanking' committed by Japanese troops in 1937 and 1938 (Chang 1997; see also Gottschall 2004 and Goldstein 2001). In addition to 300,000 (mainly men) being killed, between 20,000 and 80,000 women were raped, and often mutilated. Hence, rape was clearly widespread. However, according to Chang (1997), the rapes were not perceived as strategic by the Japanese commanders. Fearing the consequences of the rapes that fed resistance and resentment among the population, commanders introduced the (in)famous 'comfort' system whereby (mainly Japanese) women were brought in to 'satisfy the sexual needs' of the troops. Hence, the Japanese military hierarchy tried to limit, rather than encourage, rape by the introduction of military brothels (Goldstein 2001), according to the logic of the substitution narrative (which, as we will recall, is a variation of the 'Sexed' Story introduced in Chapter 1).

How, then, can variations in conceptualizations of the strategicness of rape be understood if we turn to research on the military, and warring more generally, for ulterior frameworks for understanding?[6] Is it possible to attribute variations in levels of sexual violence to certain characteristics of armed groups and units? Let us attend briefly to what some key literature has to say on this.

Understanding rape as a manifestation of greed? The fact that several armed groups (including state armed forces) often rely on civilian populations for survival (logistics, food, intelligence, etc.) is often presented as the main reason why commanders might construe rape as counterproductive (see Wood 2009, 2010; see also Kalyvas 2006 and Weinstein 2007). Needless to say, violence against civilians – including sexual violence – risks destroying the support and trust of civilian populations.

Moreover, it is often argued that assessments of the (non-)strategicness of rape depend on the (more) long-term aims of the armed group itself. Wood (2009) argues that armed groups driven by the long-term goal of governing civilians are less likely to tolerate or encourage mass rape of their future constituency (since that greatly reduces the possibility of gaining support and legitimacy). Similarly, armed groups with a strong ideological agenda, such as liberation movements, are assumed to be more inclined to limit crimes of sexual violence since they go against the norms of the new projected society (Wood 2009, 2010). This is particularly so, it is argued, in armed liberation movements, which have placed gender equality on the liberation agenda and have included women in the ranks. These movements not only tend to treat women combatants better *during* the armed struggle (see Coulter et al. 2008), but also it might be assumed that they are less likely to encourage sexual violence, fearing possible acts of revenge rape on their own troops (Wood 2009).

Wood (ibid.) has analysed several armed groups which are characterized by a relative absence of sexual violence, including the FMLN in El Salvador. She convincingly attributes the low levels of sexual violence committed by the FMLN to, on the one hand, its ideology, in particular the influence of liberation theology on new recruits, and, on the other, its dependence on, and close and cooperative relations with, civilian populations. Despite a complicated command structure, it seems as if the FMLN's ideological training produced norms against sexual violence in such a way that military commanders did not have to enforce these norms through severe punishments (ibid.).

Weinstein (2007) distinguishes between ideologically and politically motivated armed actors on the one hand, and 'opportunistic' armed groups motivated by resource extraction on the other. His research suggests that armed groups driven primarily by goals of resource extraction would have less interest in controlling violence against civilians since they are not driven by a long-term goal to govern and gain the popular support of civilians. Weinstein (ibid.; see also Wood 2009) argues that in cases where motives are primarily ideological and units depend on civilian support for survival, incentive structures work towards restraining violence against civilians. By contrast, armed groups with opportunistic motives, because of the poor availability of resources, are less dependent on civilian support and they are thereby assumed to lack the motivation to maintain cooperative and non-violent relations with civilians. They are, according to Weinstein, less likely to engage in longer-term strategies of social mobilization since 'the marginal benefits of moving quickly are much higher in resource-rich environments' (Weinstein 2007: 329).

Armed groups' access to resources is also assumed to have implications for recruitment. According to Weinstein (ibid.), armed groups with limited access to resources attract recruits by referring to future rewards and social endowment, which, in turn, leads to a more 'activist membership' with higher levels of commitment. Such commitment makes it easier for commanders to control troops and regulate the use of violence. By contrast, armed groups with ample access to resources have a tendency to recruit opportunistic and unruly recruits – individuals who are more likely to engage in violence in a short-term pursuit of economic benefits (ibid.).

While these types of distinctions between motivations of armed groups can be useful in that they illuminate the variety of warring contexts and motivations, they also risk producing a simplistic dichotomy between economic and ideological motives, which reflects poorly the complex and shifting dynamics of armed groups and armed forces (cf. Guichaoua 2012b: 3–4; Richards 1996; Utas and Jörgel 2008). As Guichaoua (2012b: 3–4) concludes:

> The 'greed/grievances alternative' has rigidified the intellectual debate on drivers of violent mobilization by reifying a bipolar conception of insurgent behaviour, at the expense of empirical accuracy and an understanding of the intrinsically fluid nature of the phenomenon of violent engagement. [...] All violent groups do, in practice, reflect a very varied mixture of intertwined motives and intentions, conscious or not, including, among others, opportunistic and ideologically driven ones.

Moreover, while a differentiation between motives surely underscores the fact that contexts of and motivations for engaging in war are different, these motivations in themselves *do not determine* the meanings attached to military strategy and tactics. In addition, it has to be recognized that the repertoire of resource extraction employed by many armed groups is quite diverse. Social and economic resources are not only 'likely to be mobilized simultaneously, but they might also be symbiotically intertwined' (Guichaoua 2012a: 273). This, as we will demonstrate below, is very clear in the case of the DRC.

A problematization of rape/greed: the case of the Congolese armed forces As is well documented in several reports, members of the Congolese armed forces, in collaboration with other armed and non-armed groups, are heavily involved in the illicit trade in minerals (cf. Global Witness 2009; UN Group of Experts 2004–11). However, while often the most lucrative (depending on prices on the world market), mining is but one source of income. Others include charcoal production, poaching, illegal arrests for extortion purposes, providing private security and, more generally, unofficial taxation of important business sites, routes, borders and ports, hemp production, etc. (see UN Group of Experts 2004–11; Verweijen 2013a; Laudati 2012). The driving forces and aspirations of armed groups to be integrated into the army are intimately linked to attempts to harvest the fruits of a militarized economy – but now in the more legitimate guise of state security forces, labels and uniforms.[7] Hence, the Congolese armed forces can be seen as largely driven by the motives of resource extraction. Following on from this supposition, it could be argued that civilian protection is counterproductive to its main functions. In order for the army to produce insecurity to encourage people to solicit protection and in order to enable resource extraction, abuses against civilians can be seen as useful for the army in certain contexts (Verweijen 2013a). This reasoning is in line with the argument, discussed above, that armed groups with so-called opportunistic motives lack incentives to treat civilians well (Weinstein 2007).[8]

However, concluding that rape and 'other' violence *is* a consequence of resource extraction is to obscure a much more complex reality. In fact, in many cases, effective resource extraction necessitates good relations with local authorities and communities, something that is difficult to create and maintain with high levels of abuses.[9] The same armed group or unit can be involved in various activities, including: the imposition of unofficial fees and taxes, the timber trade, poaching, cattle herding, the mineral trade, etc. While some of the activities in this varied repertoire are less dependent on close and non-violent relations with civilian populations and authorities, some require closer cooperation with local authorities and populations in order to be effective (Verweijen 2013a; see also Laudati 2012; Titeca 2011). In order to engage in more long-term resource extraction, the Congolese armed forces (like other armed groups) often have to create coalitions with local authorities who enjoy some kind of legitimacy, such as customary or administrative leaders or religious authorities (Verweijen 2013a). Hence, in these instances, violence against civilians, including sexual violence, is counterproductive for effective resource extraction.

In sum, violence against civilians, including sexual violence, can be understood as *both* strategic and non-strategic for the Congolese army, depending on the context. That the army may act in an opportunistic manner does not mean that this opportunism necessitates or is aided by inflicting violence against civilians. Moreover, it could be argued that the continual redeployment of units (rather than resource extraction or opportunism in itself) feeds into higher levels of abuses, since it reduces the possibilities to create more long-term profitable relations with local authorities, rendering units more inclined to short-term, and more violent, forms of resource extraction (see also ibid.). In addition, it has to be recognized that prevailing discourses and attendant norms of behaviour vary from one military unit to another in this quite diverse army. Some commanders and units are more tolerant of abuses or more likely to adopt violent strategies than others (ibid.; Eriksson Baaz and Stern 2010).

In sum, the discourse of Rape as a Weapon of War is simplistic in that it tends to write sexual violence as inherently strategic in militarized contexts. Learning from empirically based research on such militarized contexts, we can see how military discourses diverge and military commanders not only encourage but also try to curb sexual violence (see Wood 2009, 2010). As Goldstein (2001: 368) concludes in his historical overview of gender and war, '[c]ommanders' attitudes range, in various ways, from relative tolerance or even encouragement, to relatively strict punishments of rape'. Our picture of rape as a 'strategy' becomes much more complicated and nuanced than that presented by the generalized story offered by the Rape as a Weapon of War discourse.

Let us now turn to exploring a further silence/misrepresentation in the discourse of Rape as a Weapon of War: namely the failures of military actors to live up to their celebrated concepts of discipline, control and order.

Failures of military institutions to embody discipline and control

One central supposition in the discourse of Rape as a Weapon of War seems to be that military institutions operate in orderly and efficient ways. The occurrence and pervasiveness of rape in warring contexts appear as a sufficient testimony that sexual violence is used strategically by military commanders (see Chapter 2) (cf. Seifert 1996; Skjelsbaek 2001). Consequently, military contexts are portrayed as working in precise ways, with orders effectively enforced down the chain of command by disciplined military staff. If this were the case, 'sexual violence in the war-zone' would indeed be 'simply too widespread [...] for it *not* to be part of a larger political scheme and hence a weapon of war', as suggested by Skjelsbaek (2001: 213, quoted in Chapter 2). However, as we will argue here, military institutions very rarely operate in this orderly way. Instead of reflecting the tidy workings of the military, the widespread occurrence of sexual violence, we suggest, can also be seen to reflect the breakdown of the chain of command; indiscipline instead of discipline, unrestrained fear instead of control.

Searching to attain the impossible The formal importance attached to training and the informal importance attached to initiation rites within military institutions is itself a lucid acknowledgement of the fact that even military institutions themselves understand that military ideals do not come naturally or easily (cf. Goldstein 2001; Higate 2012b; Soeters et al. 2006; Cortright 1975). Paul Higate (2012b) reflects on Enloe's (2004) classification of militarism, and notes that the idea that 'human nature is prone to conflict' as a core militaristic value tends to 'ignore the workings of military institutions [...]. If it is indeed the case that human nature is prone to conflict,' he argues, 'then why the concerted, intensive and hugely expensive time, effort and resources dedicated to training civilians to become soldiers?'[10]

Through (various degrees and forms of) training and initiation rites, recruits learn how to become a military person who embodies the values of hierarchy, discipline and control.[11] The generalized 'Gendered' Story of military masculinity presented in Chapter 1 builds upon this reasoning, but places its focus on how this 'person' is necessarily masculine, and how militarization involves the production of violent masculinities and, ultimately, rapists. Here, instead, the emphasis is on the work involved in producing soldiers and the military hierarchical structure and institution, in part as a mechanism to carefully govern the use of violence. As Soeters et al. (2006: 250) conclude, the military aims to be (in Goffman's term) a 'total institution' and recruits are supposed to undergo a process of 'degradation or "mortification", i.e. a process of deconstruction of their civilian status'. After that, 'having become receptive to new values, the cadet-officers are "rebuilt", i.e. given a new identity' (ibid.). Hence, great effort is directed at creating committed and disciplined soldiers, who will 'engage in consistent and predictable behaviour' (ibid.: 250). The resources and energy dedicated to this process reveal that it is not understood as a simple process by military establishments. Moreover, the 'greedy' (ibid.) nature of the military institution (i.e. its high demands on its personnel and its reliance on a strict disciplinary framework, sometimes involving severe punishments such as the death penalty for desertions) shows that this process is incomplete. Despite repetition through continual training, drills and punishments for transgressions, the aspired-to ideals are never fully attained.

This 'incompleteness' is often not sufficiently recognized in the Weapon of War discourse, or in feminist research on militarization more generally. Soldiers tend to be portrayed as obedient passive subjects of training, which is geared towards moulding reliable masculinized killers (and rapists). Hence, there is a tendency in this literature to downplay the agency of soldiers themselves, who are not simply passive recipients of training, but engage in various forms of coping strategies and resistance (Grossman 2009;

Hockey 1986; Kirke 2010). By not sufficiently recognizing this agency, there is a tendency to overstate the capacity of the military institution to create and govern obedient subjects and the use of violence. Moreover, as Kirke's (2010) ethnographic study of the British military reveals, certain levels of 'rule bending' and 'rule breaking' – rather than complete discipline and obedience – can also be understood as an integral aspect of socialization processes in the military, and as embedded in the organizational culture more generally (see also Hockey 1986).

In addition, it must be recognized that training and preparation are certainly quite different from real-life experiences of war, despite any 'realism' training employed (see also Bourke 1999). The history of warring is crowded with examples illustrating the incompleteness of the ideals of discipline and control. As Goldstein (2001: 253) concludes, 'war is hell' and '[a]ny sane person, male or female, who is surrounded by the terrifying and surreal sights and sounds of battle, instinctually wants to run away, or hunker down and freeze up'. In these situations, the rehearsed tasks and common codes of understanding sometimes collapse, and give way to panic, confusion, misinterpretation of orders, disobedience and even desertion.[12] Simply put, we would do well to recognize that warring is characterized by 'uncertainty' and 'the chaotic and unpredictable unmaking of certainties' (Barkawi and Brighton 2011: 139). As Shane Brighton (2011) argues in a reading of Clausewitz on the phenomenology of war,[13] war presents itself as:

> a field of contingency in which unpredictability and the general absence of certainty dominate and 'the light of reason is refracted in a manner quite different from that which is normal in academic speculation' [...]. Outcomes cannot be predicted with certainty, assumptions are violently unmade and new ones generated. So often the bonfire of certitudes, war disrupts the claims of foundational thinking. As a process of violent reciprocation, 'Clausewitzian war' is always to some extent beyond conceptual capture, always a field of uncertainty, always potentially in excess of the attempt to fully command it. (Ibid.: 102)

Let us now venture into the more concrete: into how literature within the field of military sociology accounts for the role of military cohesion in the effective enforcement of orders down the chain of command and effective warring.

Disintegration of vertical cohesion and diverting pent-up frustrations One of the most intricate tasks is to create and maintain social cohesion in military units (King 2007; Siebold 2007). One central aspect of this is so-called 'vertical cohesion', the bonding/linkage between unit members and their leaders (in contrast to horizontal cohesion, or peer bonding, signifying bonding between soldiers at the same level in the military hierarchy) (see Bass et al. 2003; Siebold 2007). While there are different views on how to best define military cohesion, as well as how to achieve it (cf. Bartone et al. 2002; King 2007; MacCoun et al. 2006; Siebold 2007; Wong 2006), there is a general consensus on its importance for military performance. The vital importance of vertical cohesion is also reflected in the emphasis put on leadership skills in military officer academies. Officer training aims to produce leaders who are capable of creating and maintaining the loyalty of troops; officers who embody the virtues of loyalty, honesty, empathy and caring (about their troops), while able to rapidly make informed decisions in times of crisis in ways that do not put the lives of their subordinates at too much risk (see Caforio 2006).

Indiscipline and the breakdown of vertical cohesion take various forms, from minor day-to-day resistance and evasion, to more violent forms. The best-studied instance of widespread disintegration of command structures occurred during the Vietnam War, when large numbers of officers were deliberately killed by their subordinates through fragmentation grenades (giving name to the term 'fragging'). While the Vietnam War was accompanied by very high levels of documented assaults and killings of superior officers, violent attacks against commanders are by no means exclusive to the US Army, but are instead a well-known phenomenon throughout the history of warring (Cortright 1975). Most often, the breakdown of discipline and vertical cohesion occurs in challenging and gruelling combat situations and is coupled with subordinates' assessments that their commanding officer is incompetent and exposes them to unnecessary risks. Sometimes it has more of a 'class aspect' and is associated with the dissimilar conditions of officers and soldiers alongside the sentiments of the latter that they unjustly endure all the hardships of combat while officers lead a privileged life (ibid.).[14] Moreover, the breakdown of the command chain and assaults on senior officers also tend to occur in situations where troops start to question whether they are fighting for a just cause (as was the case in the Vietnam War).

This venture into exploring possible reasons for the disintegration of military hierarchies helps us to argue from yet another angle, namely that the widespread occurrence of sexual violence (as with other forms of violence against civilians) does not necessarily imply that it is a conscious strategy encouraged by commanders. Sexual violence can also reflect the breakdown of chains of command; indiscipline rather than discipline; commanders' lack of control rather than their power.

A central aspect of 'dysfunctional' chains of command is the limited and distorted flow of information. Hence, even when information about non-sanctioned violence against civilians (such as rape) reaches commanders, smaller military units with strong group solidarity can choose to withhold and distort information, making it very difficult for commanders to identify and punish perpetrators (see Wood 2009, 2010; Goldstein 2001). Moreover, even in cases where commanders are made aware of a certain act of indiscipline, they can be too afraid to impose sanctions, fearing it may trigger an uprising against them. Following this reasoning, allowing soldiers' frustration to spill over/be channelled through non-authorized violence against civilians is a safer route when sanctions risk producing a situation in which frustrations are directed towards superiors. (This reasoning presupposes, of course, that one believes that frustrations are often 'channelled' into non-authorized violence. We will return to a discussion of research that supports this supposition in the following section.)

In addition, it must be remembered that the chain of command and patterns of sanctions are not static in any given army or armed unit, but shift according to context. For example, while commanders in the German Army applied severe punishments to soldiers committing the crimes of rape and pillage on the Western Front, they largely let the same crimes go unpunished at the Eastern Front, which was characterized by extremely harsh conditions for the soldiers. As Goldstein argues, on the Eastern Front, loss of self-control and violence against civilians served as a psychological 'compensation', a 'safety valve', 'diverting soldiers' pent-up anger and frustration against defenceless enemies instead of their own superiors' (Goldstein 2001: 368). As we will suggest below, the same logic is commonplace in the DRC warscape and sheds light on some of the sexual violence committed.

Disintegration of vertical cohesion and diverting pent-up frustrations: the case of the DRC The Congolese army provides ample examples of the ways in which state military institutions can fail to operate in their 'ideal' sense.[15] Violence against civilians, including sexual violence, occurs in the context of these failures; consequently, understanding sexual violence necessitates interrogating how such failures facilitate and are even conducive to the propagating of conflict-related rape.

First, large parts of Congolese army units are characterized by quite meagre vertical cohesion.[16] This can be attributed to various factors. In addition to the challenges of building units from former adversaries and making these units unite behind one commander, the constant reorganization and reshuffling of military leaders undermines cohesion. Moreover, the situation is exacerbated by the generous policies of integration (Eriksson Baaz and Verweijen 2013), which created situations where the newly integrated with little or no formal military training but high formal ranks and functions are supposed to command staff with higher level of training but lower ranks (Eriksson Baaz and Stern 2008).[17]

However, one of the main problems is connected to the general sentiment among troops that commanders are unfair and benefit themselves at the expense of subordinates. This is partly related to the real (and imagined)[18] embezzlement of soldiers' salaries by commanders, but also to perceptions of the unfair distribution of the unofficial resources soldiers collect for commanders. Most interviews with soldiers featured deep-seated feelings of neglect as well as frustrations and dissatisfactions with superiors (see ibid.). As one soldier put it:

> A good soldier is a soldier who follows *Règlement Militaire*, who has discipline and obeys it. But how can we do a good job when we do not have anything. Here [in Kinshasa] we are hungry and at the front we are hungry. We don't get anything. They cheat us. [...] According to the rules we are supposed to get rations, food, medical care, but now there is nothing. [...] So tell me, how can we be disciplined? They all cheat us. Our superiors cheat us. We die and our children die. They send their children to Europe, but our children die.[19]

Taken together, these factors inhibit adequate levels of vertical cohesion. This situation, we argue, may contribute to the occurrence of sexual violence in various ways. First, it makes it difficult for the commanders who try to limit violence against civilians to effectively enforce the rules. Secondly, it reduces the flow of information to military commanders, a problem that in the DRC is amplified by the fact that military units often operate far away from their commanding officers for long periods of time without access to communication equipment (Verweijen 2013a). Thirdly, low levels of vertical cohesion reduce the incentives and willingness of commanders to hold abusers to account. Widespread levels of frustration among troops can mean that some commanders live in more or less constant fear of rebellion. In turn, this renders them less inclined to punish indiscipline and non-authorized violence against civilians. In keeping with many other warring contexts, the tolerance for violence against civilians can be understood as a way to divert pent-up frustrations. Moreover, it could be argued that a permissive climate and reluctance to punish violence against civilians, particularly in relation to property violations and rape, work as informal compensation for the lack of formal salaries and benefits. We can recognize the familiar 'war booty' reasoning found in the 'Sexed' Story, appearing here as an explanation for why commanders 'allow' sexual violence. Hence, these violations could still be configured as 'strategic'. However, as noted above, the strategicness here is different from that implied in the Weapon of War discourse in that it is located not in its effectiveness in humiliating and punishing civilians, but in its usefulness in protecting commanders' power positions.

Last, a further aspect that feeds violence against civilians is the army's weakness as a combat organization. As Verweijen (ibid.) demonstrates, the Congolese army often operates with a set of loosely defined instructions and quite limited central control, owing in part to limited communication. This means that operations are strongly shaped by localized, contingent and reactive factors, often producing disappointing results. In turn, this feeds frustrations, which sometimes find their outlet through the abuse of civilians. Seen this way, violence against civilians sometimes results from contingent events where isolated acts of violence lead to cycles of deteriorating military–civil relations, which, in turn, produce more generalized violence. As soon as someone in the unit has committed violence, support from the population diminishes sharply. Units are then more inclined to resort to intimidation and abuse in order to access resources (ibid.). While this is not particular to the Congolese army, the Congolese armed forces' (dis)organizational structure and limited resources make this turn of events more likely.

Hence, tolerating rape and other abuse can be seen as a way for commanders to divert pent-up frustration, thereby trying to mend or cover up fractured cohesion. However, Dara Cohen's research suggests that rape – particularly gang rape – can also sometimes work more directly to enhance combatant socialization (Cohen 2010, 2011). Through an analysis of different data sets from various conflicts, as well as interviews with ex-combatants in Sierra Leone, Cohen argues that armed groups that rely on abduction are more likely to commit gang rape. In such groups, rape becomes a way to 'create bonds of loyalty and esteem from these initial circumstances of fear and mistrust' (2011: 3). Cohen's analysis differs from other research, which tends to emphasize the benefits of rape for the perpetrators as well as the ways in which rape, along with looting, becomes a symbol of power and victory and acts as a source of reward (again, the familiar 'war booty' argument). Cohen, in contrast, focuses on the 'flip side of the costs and benefits' (ibid.: 12) and argues that rape entails various grave personal risks and costs for the individual, such as health risks, 'the singular emotional toll of the intimate contact required' (ibid.: 12), and the fact that rape is a relatively time-consuming crime. By contrast, she argues, gang rape provides abundant benefits for the group and it is the personal risks involved (rather than benefits) which, according to her, reinforce the workings and utility of group rape as a mechanism of cohesion. Shared personal risk binds the group together. Hence, Cohen characterizes the group rape in her studies as a bottom-up process, and not a bottom-down military strategy or tactic.[20] She argues:

> Unlike explanations focused on private motivations, an argument based on combat socialization does not depend on either the combatants having a biological or latent desire to rape non-combatants, nor the idea that rape must have an overtly military purpose. The argument merely posits that, conditional on being trapped in a group of hostile strangers, individuals often choose participation in costly group behaviour over continued estrangement from his or her peers. (Ibid.: 12)

While the above discussion offers well-grounded (in our research as well as others') and helpful analysis, these lines of argument do not address the question of why and how (sexual) violence appears as a solution, an opportunity or a necessity in the context of a 'dysfunctional' military structure. To further explore why violence becomes possible, let us turn to the last of our identified 'gaps' in the discourse of Rape as a Weapon of War, namely the micro-dynamics of war and the workings of cycles of violence. In the above paragraphs, we have contended that violence in war cannot simply be understood as stemming from orders from the military or political hierarchy. Below we will attend to some of the other dynamics that may shape patterns of violence in war. If violence is not always ordered, and if the 'sexual urge' proposition (at least in its crude form) offers little explanation, how else can we understand the dynamics of sexual violence committed in war?

The micro-dynamics of violence in war

Forward panic and spirals of violence Randall Collins's (2008) micro-sociological analysis of violence in and outside of war problematizes the 'rationality' of violence. His analysis of the concept/practice of violence is general and includes all forms of violent acts, including rape, particularly violent rapes featuring mutilations and/or the subsequent killing of the victims. 'Forward panic' is a central concept in Collins's analysis, which can be understood as an emotional state feeding a frenzy of excessive and non-utilitarian violence. Collins describes forward panic as an emotional flow that can arise in circumstances of intense tension or fear. Forward panic occurs when such situations transform 'into a sudden rush of frenzied overkill in an atmosphere of hysterical entrainment' (ibid.: 100). Through an in-depth reading of perpetrators' experiences of violence, he provides examples from several atrocities connected to 'forward panic', including the My Lai massacre in Vietnam in 1968 (as well as other atrocities during the Vietnam War), Nanking, several ancient battles such as the battle of Thapsus (46 BC), but also non-war atrocities such as the beating of Rodney King and other instances of violent police intervention and crowd violence.[21]

Collins describes forward panic as a state characterized by 'hot emotion' (being 'highly aroused' or 'steamed up') and a 'rhythmic entraining emotion' (ibid.: 93), where those affected repeat violent actions in a frenzied manner. Collins exemplifies this with the familiar tendency of front-line troops to kill enemy soldiers who try to surrender. While acknowledging that such acts can also sometimes be deliberate (to avoid the burden of taking and caring for prisoners) he argues that the common killing of surrendering enemy troops in front-line areas can also reflect the workings of the state of forward panic. While the act of surrender signals victory, front-line troops (in a state of forward panic) are often caught in 'the situational momentum' (ibid.: 95) and simply continue the repetitive violent acts of the battle. According to Collins:

> A forward panic is violence that for the time being is unstoppable. It is over-kill, the overuse of force far beyond what would have been needed to bring about the victory. Persons who have fallen off the point of tension into a forward panic situation have gone down into a tunnel and cannot stop their momentum. They fire far more bullets than they needed; they not only kill but destroy everything in their sight; they throw more punches and kicks; they attack dead bodies. (Ibid.: 94)

One US platoon commander in Vietnam described his experience of a destructive frenzy (Collins describes this as an example of forward panic):

> The noise of the battle was constant and maddening, as maddening as the barbed hedges and the heat of the fire raging just behind us [...] Then it happened. The platoon exploded. It was a collective emotional detonation of men who had been pushed to the extremity of endurance. I lost control of them and even of myself. Desperate to get to the hill, we rampaged through the rest of the village, whooping like savages, torching thatch huts, tossing grenades into the cement houses we could not burn. In our frenzy, we crashed through the hedgerows without feeling the stabs of the thorns. We did not feel anything. We were past feeling anything for ourselves, let alone for others. (Caputo 1977: 287–9, cited in ibid.: 87)

Collins emphasizes that there are multiple paths to atrocities and that 'forward panic' is merely one among others. He distinguishes between atrocities in war ensuing from 'forward panic' and atrocities emanating from 'deliberate orders from high military or political authorities', 'scorched earth policy' and 'exemplary punishments' (ibid.: 99–100). The central characteristic of violence arising from 'forward panic' is that it is non-utilitarian, situational and local. However, while emphasizing the distinctiveness of 'forward panic', he also argues that the various pathways of violence can overlap and that some acts of violence must be understood as manifestations of a confluence of various routes. This was, according to him, the case in Nanking, where orders by the military hierarchy to kill all the prisoners escalated into an 'orgy of destruction' (ibid.: 99), including killings, rape, mutilations and looting, leaving commanders unable to control the troops.[22]

Other researchers demonstrate the workings of spirals of violence and how those who feel humiliated, mistreated and victimized by the enemy (or even through the context of warring more generally) tend to become more prone to enact violence against perceived enemy populations, soldiers as well as civilians (Horwood et al. 2007; Kassimeris 2006; Weiner 2006). Violence loses its taboo; the more violence one witnesses, suffers from or inflicts, the easier it is to become 'morally disengaged' from those whom one sets out to harm and torture (Muñoz-Rojas and Frésard 2004; Staub 1989, Zimbardo 2008). Facets of such a spiral include perpetrators viewing themselves as victims, finding 'justification' for violent behaviour (i.e. they 'deserve' it and therewith it is 'right' to seek revenge), shifting blame away from oneself, and distancing oneself from one's victims through processes of Othering (Muñoz-Rojas and Frésard 2004; Staub 1989). Furthermore, the prevalent use of drugs and alcohol in many warring contexts enables the perpetrator to feel even more removed from a sense of agency and responsibility (Collins 2008; Kassimeris 2006; Kovitz 2003).

Testimonies from soldiers in various war zones bear witness to the workings of both forward panic and spirals of violence. Soldiers and combatants often testify to how they, within a quite short period of time, become perpetrators of violence they never imagined they could commit. Stories told by soldiers in Nanking, for example, convey experiences of this process. Most soldiers told of experiencing complete shock at the extreme violence (not only rape) when they first arrived, but how they shortly became part of such violence themselves. As one soldier explained, his initial experience of the violence was 'so appalling that I felt I couldn't breathe', but that '[e]veryone became a demon within three months' (Chang 1997, cited in Goldstein 2001: 36).

According to Collins, a central feature of forward panic is that it is experienced as an 'altered state of consciousness, from which the perpetrators often emerge at the end as if returning from an alien self' (Collins 2008: 100). The US platoon commander in Vietnam (cited earlier) explained his experience of the frenzy of destruction described above in the following way:

Of all the ugly sights I saw in Vietnam, that was one of the ugliest: the sudden disintegration of my platoon from a group of disciplined soldiers into an incendiary mob. The platoon snapped out of its madness almost immediately. Our heads cleared as soon as we escaped from the village into the clear air on the top of the hill [...] The change in us, from disciplined soldiers to unrestrained savages and back to soldiers, had been so swift and profound as to lend a dreamlike quality to the last part of the battle. Despite the evidence to the contrary, some of us had a difficult time believing that we were the ones who had caused all that destruction. (Caputo 1977: 287–9, cited in ibid.: 87)

Let us now turn to how cycles of violence and forward panic discussed above can be seen to be manifested in the DRC case and how that might provide us with a more informed understanding of the occurrence of sexual violence.

Forward panic and spirals of violence: the DRC The interviews we conducted with soldiers in the DRC bore witness to strong feelings of humiliation, mistreatment and victimization – sentiments that, arguably, make people more prone to committing violence (Horwood et al. 2007; Kassimeris 2006; Weiner 2006). These sentiments were articulated not only in relation to the enemy or the military hierarchy but to civilians in general. Civil–military relations in the DRC are highly ambivalent and contextual and sometimes less hostile than often portrayed (see Verweijen 2013a). Nonetheless, the long history of abuse perpetrated by the military (dating back to the colonial era), in combination with their low status (which is connected to low salaries, poor living conditions and the lack of non-material rewards, e.g. medals, as well as nationalist propaganda in support of the armed forces), has perpetuated a negative image of military personnel among civilians. This recurs in the soldiers' accounts (although less so in interviews with officers, who enjoy more respect in light of their position). The soldiers repeatedly underscored the extent to which they were misunderstood and disrespected by civilians. In many testimonies, violence against civilians was clearly expressed as a manifestation of a 'need to put them in their place', 'show them a lesson' and 'punish' them. As one corporal explained:

The civilians don't respect us. They see us as useless people/losers [*batu ya pamba*]. Because we don't have anything. We have to beg from them, so they see us as losers. They call us bad names [*bazali kofinga biso*] [...] Yesterday when I was out, somebody spat on me. Sometimes they even attack us. We don't go out alone any more, not alone without any weapon. That can be dangerous, because sometimes they can attack you. Here, in this area, there are a lot of weapons around. The civilians have lots of weapons and they can kill you [...] So therefore, sometimes you have to show them [*par fois il faut olakisa bango po pe bayeba*]. They are thick headed [*bazali mutu makasi*]. They don't understand things. So sometimes they need some punishment [...] That is also sometimes, sometimes, an explanation for rape. If they respected us, it would be different. Then you would not see so much of all that, rape, killings and stealing. It is also that. Their disrespect [*manque ya respect*]. They don't understand.[23]

While the wording here (and in many other accounts) is similar to that used in the discourse of Rape as a Weapon of War (e.g. 'punish'), these acts of punishment were not discursively linked to (i.e. not described as following from) strategic goals or orders by commanders. Instead, they were described as linked to sentiments of humiliation and mistreatment. As demonstrated in this quotation, soldiers portrayed themselves as victims, and it is in this position of 'victimcy' (Utas 2005) that they found the justification for their violent behaviour. They shifted the blame from themselves and distanced themselves from the civilian population through processes of Othering. While, as explored above, there is an aspect of strategicness in some of this violence in that it may work to divert pent-up frustrations from being enacted against the military hierarchy, this violence is not mainly connected to military goals. Instead it seems more likely that it reflects cycles of violence and processes of Othering, connected to the context of warring and conflicting civil–military relations in many deployment areas. (Feminist analysis of rape as a performative act of power and domination would certainly help to further deepen this point, as the sense of disempowerment and frustration evident in the soldiers' accounts appears as an attempt to regain a semblance of power and 'respect'.)[24]

Moreover, many soldiers testified to the ways in which they became familiar with and dulled by the effects of violence, as well as how they distanced themselves from those they set out to harm. Some, especially those who had previously been engaged in armed groups, explicitly described how they had committed violence against civilians as part of their initiation into the violence inherent in fighting. As one soldier put it:

In the beginning it is difficult [*pasi*]. The first time it is difficult. That is scary and you feel really bad. That is also why they force you in the beginning. Me, they told me to kill a prisoner. It was really difficult. They do it to get the civilian spirit out of you. You just have to do it. Because you have to do it. Otherwise they will punish you – or even kill you. But after a while you get more used to it. But in the beginning it is really hard, really, really hard.

Some further explained that after this initial transformation into someone who could/would kill, they then continued to engage in the violence of warring with little trepidation.

Numerous soldiers talked about the 'spirit and craziness of war' and the use of drugs as explanations for the violence committed against civilian populations. Many of these narratives described violence, including rape, as a result of a 'spirit of war' which makes people 'go crazy', alluding to how warring is an unnatural and extreme state. Here we recognize what Collins (2008) described as 'forward panic', as well as the seeming (re)solution of the familiar rational/animalistic dichotomy; warring temporarily transforms the rational subject into a 'beast', who then retreats so that the rational subject can reflect on his 'bestial' state (see above). Rape, like killings and looting, was often described as resulting from anger and frustrations that are manifested in a general urge to destroy. One captain described the effects of war in the following way:

> War is crazy, it destroys the minds of people [*ezali kobebisa mitu ya bato*]. Some people just go crazy [*bakomi liboma*]. Rape is a result of that too, especially the bad rapes. It gets too much [...] Also, a lot is because of drugs. If you take drugs, drink, or other things – it is not good. And many, many [...] most take drugs.

As reflected in this quotation, the use of drugs occupied an important position in the narratives. Drugs were described as a necessary way to cope with the hardships of life as a soldier (poor living conditions, etc.), but particularly in order to cope with the fears of combat situations (see Eriksson Baaz and Stern 2010; see also Kovitz 2003).[25]

In reading the accounts above, we recognize the assumed rational subject who acts in the story of rape told through the Rape as a Weapon of War discourse. In this storyline, this rational subject reverts back to/falls into savagery, does terrible incomprehensible things, and then re-emerges as his 'normal' self. The savagery depicted here seems to inhabit and overcome the rational human self, and is unleashed in the violent climate of warring, only to retreat again. This framework thus makes room for both the 'beast' and the rational subject (albeit not at the same time), thus disrupting slightly the simple dichotomy presented in reductionist renderings of the 'Gendered' Story, while still upholding its basic lines of distinction.

The irresistible opportunities of war: violent private score-settling and blurred boundaries between military and civilian spheres Another strand of research problematizes the notion of violence in war as stemming from rational calculation by military and political leaders by pointing to the fluidity of military and civilian spaces and the distinctions between them during war. As Kalyvas (2006) demonstrates, there has been a tendency in research on civil wars to neglect the fact that much violence is related to micro-level dynamics and private disputes, involving both armed and non-armed actors. As he concludes, civil war 'politicizes private life' (while at the same time 'privatizing politics'), and 'offers irresistible opportunities to harm every-day enemies' (ibid.: 389).

'Master narratives' of wars (with their macro-level stories of who the pro-tagonists are understood to be and what the war is all about) are often framed by simplistic distinctions between combatants/soldiers and civilians. These narratives frequently offer a one-dimensional reading of violence in war as imposed by armed actors on innocent civilians, and ignore the ways in which much violence in war is a manifestation of micro-dynamics, personal disputes and score-settling. Hence, rather than simply reflecting decisions of military and political leaders, violence in war contexts, including sexual violence, must also be understood in relation to the opportunities that war offers in providing violent solutions to already existing or emerging local conflicts and animosities (ibid.). As we will discuss below, such dynamics occur often in the DRC, with its marked blurring of civilian and military lives, where soldiers live within and become implicated in the settlement of personal scores within the civilian communities.

Some of the violence committed by the Congolese army can be understood in relation to the ways in which the army intervenes to assist in settling personal scores within civilian communities. This role is facilitated by the fact that Congolese army units live among civilians in many places (rather than in military camps), which increases their involvement in different aspects of 'civilian affairs'. One of the income-generating activities of the army (enacted both collectively and by individual soldiers) is to provide their services to settle various kinds of private and family conflicts, often through violent means. While this violence mostly involves intimidation, the beating up of adversaries, property destruction or seizure, and assassinations, etc. (see Verweijen 2013a), it can be assumed that the 'assistance' provided in violent score-settling and the often ensuing spirals of violence also contribute to sexual violence.

Moreover, some 'mass rapes' should be seen as an effect of private disputes that turn into spirals of mass violence; the rapes, lootings and killings in Fizi in January 2011, for example, can be understood in this way.[26] These acts of violence were committed in a context of already tense relations between the army unit deployed and civilians, and were triggered by an argument (allegedly over a woman) between a soldier and a civilian in a bar at night.

The argument escalated into a fight in which the soldier shot and killed the civilian and, in response, a lynch mob comprised of civilians then killed the soldier. A revenge attack followed in which soldiers, encouraged by their commander, retaliated further.[27] While it is probable that these rapes were indeed sanctioned by the commander (who was later sentenced to twenty years in prison; BBC 2011b), this example underscores the workings of cycles of violence and the ways in which sexual violence – like 'other' forms of violence – is often a manifestation of unplanned, spur-of-the-moment acts. Rather than functioning as premeditated strategies to punish civilians in order to achieve some military goal (whether in the form of combating the enemy or resource extraction), such spontaneous violence feeds into cycles of violent score-settlement.

Concluding discussion

We have discussed in this chapter how military institutions often fail to operate according to the ideals of discipline, hierarchy and control. Relatedly, the occurrence of sexual violence in warring contexts does not necessarily imply that it is promoted or construed as strategic by the military leadership. One of the predicaments of the discourse of Rape as a Weapon of War is that it does not seem to acknowledge these 'failures', but instead portrays the military institution as the rational war machine it aspires to be. Consequently, and ironically, the Weapon of War discourse thereby tends to reproduce the flattering self-image of the military propagated by the very militaristic views that it rejects.

We would like to emphasize that we, in no way, contest the fact that sexual violence in certain contexts is construed as strategic and encouraged by military commanders.[28] Instead, we argue that if we are to be able to grasp the complexity of wartime rape in order to better remedy its occurrence, we would do well to pay close attention to research and experience that indicate that it can *also* be the opposite. Instead of reflecting the orderly workings of the military, the widespread occurrence of sexual violence can instead reflect the collapse of military hierarchies and cohesion, as is arguably often the case in the Congolese armed forces.

The discussion above, particularly soldiers' narratives of their experiences of war, clearly troubles notions of rape in war as a strategic weapon. While violence in these narratives still emerges largely as a manifestation of efforts to humiliate and punish, the underpinnings of these endeavours are much less strategic and far more complex than a combat strategy to further military objectives (however defined). The stories highlight the ways in which the logics of violence in war are sometimes far removed from the main (meta-)narratives of the war and instead are manifestations of mundane everyday personal animosities, of moral disengagement, 'forward panic' and the soothing numbness of intoxication, making people forget, while at the same time facilitating new acts of violence. The narratives, particularly the representations of the soldering self as an 'abject hero' (Utas 2008) and the attendant discursive strategies of 'victimcy' (Utas 2005),[29] are surely problematic and warrant further scrutiny (see Chapter 1). Yet their stories have something to teach us about the messiness of warring – a messiness which fortunate outsiders (including ourselves) tend to downplay and forget.

Importantly, however, the discussion in this chapter does not offer an adequate answer to the question of why violence becomes sexualized. While parts of the problematization offered above (such as Collins's concept of 'forward panic') explicitly refer to all forms of violence, including sexual violence, the question 'but why rape?' still lingers on. At the same time, the very framing of this question (of why violence becomes sexualized) builds on a somewhat problematic notion that sexual violence is fundamentally different from 'other' violence; that we *know* what violence is sexualized and what such sexualization means to the subjects of violence (the perpetrator as well as the victim); that sexual violence follows a particular logic which is inherently different from 'other' forms of violence; that we know what the logics of sexual violence are, as well as what the logics of 'other' forms of violence are. Simple answers asserting the definitional distinction between 'sexual' violence and 'other' violence risk reproducing rigid, shallow and perhaps even misleading knowledge claims. This is not to say that we conclude that the category of sexual violence is meaningless or ill informed; conflict-related rape obviously has devastating effects on the bodies, lives and livelihoods of the people subjected to it. Furthermore, wartime rape has particular meaning and effect in the politics of, and practice of, warring. Our point here is that its meaning and distinction from 'other forms' of violence are not as surely determined as the Rape as a Weapon of War discourse conveys.

In this chapter we discussed how the circumstances in which such acts of violence occur in war cannot solely be understood in terms of military strategy or orders from above. Importantly, the particular effectiveness of sexual violence as a means of humiliation, intimidation or punishment cannot be understood without the 'Gendered' Story (discussed in Chapter 1). In sum, while we contend that sexual violence in war must be understood in relation to contextual circumstances which also drive 'other' violence in war (circumstances discussed in this chapter), the 'sexualization' of violence, we aver, must also be understood by frameworks which take into account the power of sexgender discourses, as addressed in aspects of the 'Gendered'

Story discussed in Chapter 1. Hence, while refusing to offer an alternative explanation for rape in war, this chapter nonetheless has pointed to gaps in the Rape as a Weapon of War discourse through highlighting other viable explanatory factors linked to the set-up of military structures and the nature of warring.

4 | Post-coloniality, victimcy and humanitarian engagement: being a good global feminist?

Introduction

Five years ago, Lisa Shannon watched 'Oprah' and learned about the savage, forgotten war here in eastern Congo, played out in massacres and mass rape. That show transformed Lisa's life, costing her a good business, a beloved fiancé, and a comfortable home in Portland, Ore. – but giving her a chance to save lives in Congo. [...]

After seeing the Oprah show on the Congo war, Lisa began to read more about it, learning that it is the most lethal conflict since World War II. More than five million had already died as of the last peer-reviewed mortality estimate in 2007. Everybody told her that the atrocities continued because nobody cared. Lisa, who is now 34, was appalled and decided to show that she cared. She asked friends to sponsor her for a solo 30-mile fund-raising run for Congolese women.

That led her to establish Run for Congo Women, which has held fund-raising runs in 10 American states and three foreign countries. The money goes to support sponsorships of Congolese women through a group called Women for Women International. But in her passion, Lisa neglected the stock photo business that she and her fiancé ran together. Finally, he signaled to her that she had to choose – and she chose Congo.

One of the Congolese women ('sisters') whom Lisa sponsored with her fund-raising was Generose. Lisa's letters and monthly checks of $27 began arriving just in time.

'God sent me Lisa to release me,' Generose told me fervently, as the rain pounded the roof, and she then compared Lisa to an angel and to Jesus Christ.

Scrunching up in embarrassment in the darkened room, Lisa fended off deification. She noted that many impoverished Congolese families have taken in orphans. 'They've lost everything,' she said, 'but they take children in when they can't even feed their own properly. I've been so inspired by them. I've tried to restructure my life to emulate them [...] Technically, I had a good life before, but I wasn't very happy,' she mused. 'Now I feel I have much more of a sense of meaning.'

Maybe that's why I gravitate toward Lisa's story. In a land where so many 'responsible' leaders eschew responsibility, Lisa has gone out of her way to assume responsibility and try to make a difference. Along with an unbelievable cast of plucky Congolese survivors such as Generose, she evokes hope. (Kristof 2010a)

This is the story of Lisa Shannon, here told by Nicholas D. Kristof in an article in the *New York Times*. The story of her commitment to the raped women in the DRC can also be followed on a video (Oprah Show 2009) of the very show that propelled her to engage in the first place, and in her book *A Thousand Sisters* (Shannon 2011).

Lisa is just one of many people who have committed themselves to Congo's rape survivors. Numerous European and American journalists, activists, academics and representatives of diverse international organizations and governments have made pilgrimages to the DRC to meet and listen to survivors first hand. In addition to 'ordinary people' like Lisa, rape survivors have received visits from true celebrities on the global political and entertainment scenes. The most potent political visit was probably that of the (half-day) visit to Goma by Hillary Clinton in August 2009. During this visit, which she described as 'an incredibly emotional experience', she unveiled a US$17 million plan to fight Congo's 'stunning levels of sexual violence', which she labelled 'evil in its basest form' (Gettleman 2009b).

In keeping with the global trend of increasing celebrity activism (Richey and Ponte 2011; Turner 2004), rape survivors in the eastern Congo have also received more glamorous guests. In addition to the familiar face in African war zones, George Clooney, other Hollywood actors, such as Ben Affleck and Charlize Theron, have made recurrent visits to the eastern DRC to meet and listen to rape survivors. Furthermore, many other celebrities have lent their faces to various charity events, such as the OmniPeace 'Stamp Out Violence Against Women and Girls of The Congo', involving celebrity magazines/websites such as *Celebrity Gossip*, *Access Hollywood* and *Nicole Richie Fashion*. Some, such as Jennifer Aniston and Courtney Cox, have campaigned for Congolese women by urging electronics companies in the USA not to buy 'conflict minerals' from the Congo.[1]

In sum, many people have engaged in supporting the raped women in the Congo and much has been written on them in various media in Europe and the USA.[2] In Chapter 1, we explored the ethics, dilemmas and fears that attending to the perpetrators' stories evoke. We addressed the question of who 'we' (critical international relations scholars, feminists, those who act and feel on behalf of humanity; Jabri 2007a, 2007b) are in relation to the perpetrators of sexual violence. In this chapter we address this issue in relation to the victims of sexual violence. We provide a post-colonial reading of the global battle to alleviate the suffering of the raped women in the DRC and attend to the ethics, dilemmas and fears of engagement.

As we will argue, this engagement in many ways provides a 'dream specimen' for the post-colonial critic searching for examples of how 'the colonial [...] lives on in its "after-effects"' (Hall 1992: 248).[3] However, we also pause to reflect on whether such a critique is really fair; and in whose interests is it articulated? Is it not quite probable that the rape survivor actually prefers the ethnocentric agent driven by a quest to save her to the petulant post-colonial critic, whose nitpicking inquiry might contribute to a renewed silence?

The chapter proceeds as follows. Drawing upon post-colonial perspectives and a reading of news articles in the *Guardian* and the *New York Times* from 2003 to 2011,[4] we provide an overview of how the conflict and the problem of rape have been represented in Western media. This is followed by an analysis of the interventions intended to combat the so-called 'rape epidemic' and alleviate the plights of rape survivors. How were these interventions designed and with what consequences? In the final part of the chapter, we turn to the ethics, dilemmas and fears of engagement, which follow from our critical reading of global engagement in the problem of conflict-related rape in the DRC.

Imagining and representing the DRC war zone and its victims

The DRC is often portrayed as 'a land of violence, chaos, and avarice, perhaps beyond the comprehension of Western audiences' (Dunn 2003: 4). It is written, at the same time, as both 'unknown/unintelligible' and 'known', through its particularly dominant position in the 'colonial library' (Said 1978) as the 'heart of darkness' (Conrad 1990 [1902]). It requires no great efforts to find representations in which the conflict is portrayed as fundamentally Other, rendering the DRC the ultimate symbol of 'deviation' (Mudimbe 1994: xii). The conflict emerges as 'savage' (Kristof 2010b), a 'morass of violence' (Gettleman 2009c) and as 'messy and ragged' (Gettleman 2009d). Moreover, in line with the well-known colonial logic and the new Barbarism theory (Kaplan 1994), the violence in the Congo has often been portrayed as bizarre and inexplicable, particularly in earlier articles. 'No one – doctors, aid workers, Congolese and Western researchers – can explain exactly why this is happening' is the conclusion in a *New York Times* article (Gettleman 2007). Implicit in this portrayal lies a comparison with the more reasonable, ordered, understandable violence in what one might consider to be more civilized wars: wars, and the rapes that accompany them, that somehow make sense, unlike the seemingly nonsensical sexual violence in the Congo. In later articles, the story of inexplicability has been increasingly replaced by the narrative of rape as a strategy or tactic of war.[5]

As noted in Chapter 1, portrayals are often informed by a backwards glance to the classic colonial story of evolutionary development (cf. Dunn 2003; Eriksson Baaz 2005; Nisbet 1969). In these narratives, the DRC is depicted through the imagery of Europe's appalling, shabby and uncivilized past. For instance, John Le Carré wrote the following in the *Guardian*:

> A couple of years ago, on a brief research trip to eastern Congo, I chanced on a hillside village high above the old Belgian colonial town of Bukavu, and fancied myself for a moment transplanted to a village in plague-stricken Europe in medieval times: children, scary-eyed and brain-damaged by undernourishment, hobbling towards us, old hags of 40, teenage polio victims paddling themselves along on bits of packing case, deformed and toothless faces smiling grotesquely as they begged, young bodies scarred, broken and hideously regrown. (Le Carré 2010)

Soldiers and combatants (male) (who are by default assumed to be perpetrators of sexual violence) often appear simply as barbaric, brutal, vengeful killers and rapists, who even mutilate and eat their victims (see also the discussion on 'racialized bodies' in Chapter 1).[6] Stories of cannibalism often feature in media reporting.[7] There has been a certain obsession for militia groups that are described not only as particularly barbaric, but as particularly bizarre; reporting on the conflicts in Liberia and Sierra Leone depicted military groups in a similar manner (cf. Hoffman 2011; Moran 1995; Richards 1996). For example, in a *New York Times* article from 2007 in which the Rasta militia occupies the central stage, Rastas are described as 'a mysterious gang of dreadlocked fugitives who live deep in the forest, wear[ing] shiny tracksuits and Los Angeles Lakers jerseys and [who] are notorious for burning babies, kidnapping women and literally chopping up anybody who gets in their way'. The article concludes with a warning to the readers that the gorillas native to Congo's national parks have been 'replaced by much more savage beasts' (Gettleman 2007).[8]

The Mai-Mai armed groups, also particularly well known, rarely feature in texts without the attendant words 'ritual', 'magic' or 'superstition'. For instance, an article in the *Guardian* describes them as an 'ethnic militia, recognizable by a preference for wearing animal skins and amulets believed to give magical powers' (McGreal 2006). Through these representations, the conflict appears as fundamentally Other, as fought by odd armed groups driven by irrational aims.[9] Again we see how the familiar distinction between rational subject and barbaric non-human, which emerges through sexgender and racialized tropes, works to separate the unspeakable violence running rampant 'there' from the realm of the possible 'here'.

The ways in which outsiders have rendered survivors' testimonies have frequently been characterized by a pornography of violence (Hunt 2008; Eriksson Baaz and Stern 2010). It is only the worst cases, featuring detailed accounts of almost unthinkable violence, which seem worthy of retelling to a Western audience. Jason Stearns explains that it is as if observers try to 'outdo each other with the most barbaric gang-rape scenario' (Stearns 2009b). Stearns's observations are echoed in many testimonies, which bear witness to outside observers' fascination and proprietary claim over access to the intimate details and horrors of violence. For instance, according to an observer who accompanied the UN representative Margot Wallström in visiting the victims of the mass rapes in Walikale in 2010, the convoyed group of journalists rushed out from the helicopter upon landing in a frenzied search for victims to interview, and later compared and argued over who had been able to document the worst case.[10]

Stories of rape must, it seems, feature pictures of victims in order to attract readers. The often intimate representations of injured bodies and suffering are composed in a way that would be quite unthinkable if those depicted were survivors of sexual violence in most countries in Europe and the USA. Who would even ponder the idea of letting journalists and other visitors into a hospital ward in New York or Stockholm with women waiting for, or just recovering from, surgery for rape-induced genital injuries, and urge them to speak and retell their stories to complete strangers? In global media (and even policy) reporting, women in the West are cast as subjects 'who see', and, when they are the victims of violence themselves, they become subjects to protect from intrusive visits and representations. In contrast, Congolese women appear as different; as women who are there 'to be seen' (Sontag 2003), who do not have to be protected from reliving the traumas of rape by retelling their stories over and over again. They emerge as the visitors' 'private zoo[s]' (Trinh 1989: 82); as objects whose sufferings are there to be consumed by a Western audience. Through the notorious (and generalized) 'rape story', the Congo has, once again, become a site of European (and American) adventurism and benevolence. As we will argue below, the massive engagement in the plights of Congolese rape survivors serves as an illuminating example of the re-acting of the white wo/man's burden to 'sav[e] brown women from brown men' (Spivak 1988: 297).

Saving the Congolese woman (from the Congolese man) While the colonial project was legitimized through a general discursive construction of a barbaric, uncivilized Other to be saved by the civilizing colonial mission, this process was highly gendered (cf. McClintock 1995; Mohanty 1991, 2003; Spivak 1988). Stories of the colonized woman's brutal oppression at the hands of the colonized man often featured in narratives of the uncivilized life in the colonies. Consequently, alleviating the patriarchal oppression of the colonized woman formed part of the general civilizing mission, a task that opened up new opportunities for European women. While she, like the colonial subject, was situated at an inferior stage of human evolution in the narratives of imperialism[11] (like the beast in the wild discussed in Chapter 1) (McClintock 1995), the stories of the barbaric oppression of women in the colonies provided her with a sense of superiority and mission: to reduce the plight of her oppressed sister (Burton 1990; Syed and Ali 2011: 352).

As a number of post-colonial feminist critics (cf. Lundahl 2010; McEwan 2001; Mohanty 1991, 2003; Parpart 1995a, 1995b; Spivak 1999; Syed and Ali 2011; Trinh 1989) have argued, this self-proclaimed mission has continued through the politics and practice of global development. As Syed and Ali (2011: 357) explain: Western feminists 'roam the globe with a mission to rectify the plight of poor women and children'.

> Within what Chow terms 'a circuit of productivity that draws its capital from others' deprivation while refusing to accept its own presence as endowed', white feminist saviors continue to endorse themselves as authentic representatives of the Other women. Thus a 'mood of self congratulation as saviors of marginality' visibly runs through all the operations involved in the representation and development of the subaltern, whose voice remains conveniently ignored by her white feminist benefactor. (Ibid.: 357–8)

The engagement of women, such as Lisa Shannon and Eve Ensler,[12] lends itself to such pointed critique by post-colonial feminists. Shannon's story, as retold by Kristof, duly reflects the portrayal of the white woman as the 'savior of marginality'. Shannon is written as the heroine, occupying the central stage in the story. She is the brave saviour (in fact, even an angel and the ultimate saviour, Jesus Christ) of Generose and her raped Congolese sisters. ('"God sent me Lisa to release me", Generose told me fervently, as the rain pounded the roof, and she then compared Lisa to an angel and to Jesus Christ.') The self-congratulation is, of course, refuted by 'the savior' herself, who displays the humbleness suitable for a true heroine and instead, in a familiar manner, maintains that Generose and her sisters are the real heroines.

By referring to Generose and her sisters as the 'real heroes', the protagonist echoes a classic critique of Western civilization as an ultimately destructive development that fills people with unnatural yearnings, and seduces them from their 'true nature'.[13] In a familiar manner, 'Congo' therefore functions here as the object of Shannon's desire; as an antidote to the ills of Western (modern) civilization. Shannon presents Generose and her raped sisters as her saviours, as the people she tries to emulate ('I've been so inspired by them. I've tried to restructure my life to emulate them'). They have filled her previously empty and unhappy life with meaning. ('"Technically, I had a good life before, but I wasn't very happy," she mused. "Now I feel I have much more of a sense of meaning."') Generose and her raped sisters' lives represent the true meaningful life. They are written as strong, happy and generous, despite their sufferings.

Shannon's longings for a purer, more meaningful life echo a long tradition of romanticizing and essentializing the local, indigenous and the subaltern women in development discourse (Eriksson Baaz 2005; Kapoor 2004; Pieterse 1992a; Spivak 1999). However, like other similar representations, the reversed rhetoric does not challenge the main plot. The unequal power relations between Shannon and Generose – the 'circuit of productivity that draws its capital from others' deprivation' (Syed and Ali 2011: 357–8) – remains intact. After all, Shannon is the incontestably celebrated agent and hero in the story. The expected and fitting display of humbleness does not, in any way, challenge the main narrative. On the contrary, it works to establish Lisa Shannon as a genuine hero, exactly through her (performed) humbleness and desired unlearning of modernity through a return to the 'past'.

In addition, much commitment to alleviating the plights of the raped Congolese women bears witness to how the Raped Woman's 'voice(s) remains conveniently ignored by her white feminist benefactor' (ibid.: 358). As Spivak argues in her classic article 'Can the subaltern speak?' (1988), the problem is not that the subaltern is unable to speak but that she is denied a space to speak, and that, when she speaks, she is not listened to.[14] The raped Congolese women's voices have appeared in numerous newspaper articles, documentaries, reports (and even art pieces), and some rape survivors have also been invited to Europe to retell their stories to large audiences. Yet the listening that occurs is habitually highly selective; often only one part of the Raped Woman's multifaceted story is registered by the visitor/reader.

Our observing of and participating in many arranged encounters between visitors and rape survivors revealed a common pattern, which features also in many others' accounts:[15] the visitor does not seem to hear large parts of the account told by the woman in front of her/him; the visitor listens to the story of her rape. Other parts of her narrative are frequently systematically ignored. As also discussed in Chapter 1, one aspect of her story that goes unnoticed is the workings of other violence she has experienced: stories of pillage, of family members being killed, tortured or abducted. The visitor/spectator is there to attend to stories about the woman's rape (and perhaps acts of cannibalism), not to the killing of her son or her husband. Another and related aspect of the Raped Woman's story which somehow seems to go unregistered by the visitor/spectator is her accounts of her main worries, which, in most instances, concern her lack of basic means of survival; concerns over lack of land to cultivate, lack of clean water, concerns over her hungry and often malnourished, sickly children. These worries are frequently amplified by the workings of other aspects of violence committed while she was raped, such as pillage and the burning of fields and houses, which left her with nothing.[16] But they are also simply endemic; reflecting the widespread acute poverty created by the war and years of Western-supported 'state failure' (Trefon 2011). However, the visitor/spectator is there to listen to the story of her rape and the rape survivor's worries are simply not registered by the listener, or are simply silenced by the author of the text in which her story features.[17]

How, then, can this particular partial listening be understood? It can, we believe, partly be read as a manifestation of what we addressed in Chapter 1: namely the workings of sex/gender that normalize 'other' forms of violence and the related narratives of rape as the worst possible act of violence a woman can experience (see Helliwell 2000 for a critical discussion). In this perspective, 'the failure to hear' can be understood simply as reflecting the limits of intelligibility ('how can a woman who has been brutally raped be talking of poverty as the main plot of her victimhood'?). More to the point, it reflects the failure to unlearn 'one's privilege as one's loss' (Spivak 1988: 287; 1990: 9) or, as Spivak has explained more recently, 'learning to learn from below' (Spivak 2004). It exposes the ways in which privilege and racist prejudice limit our comprehension and prevent us from gaining other knowledge and grasping other experiences. It also reveals how we project ourselves and our 'world onto the Other' (Spivak 2004, quoted in Kapoor 2004: 642).

The failure to listen is surely also a manifestation of our fascination with the spectacular. It is a reflection of the process by which we produce Congolese raped women to suit our own images and desires. The reason people visit the Panzi or Heal Africa hospitals in eastern Congo or come to the seminar in Europe to which the rape survivor (or the 'rape expert') is invited to speak is, after all, not to hear about poverty. It is to listen to stories of rape. Put alongside *wartime* rape (everyday rape appears also as somehow ordinary – even boring), poverty appears quite banal, and unable to capture our fascination. Images and stories of poverty are archaic and mundane, already subsumed as a regrettable but not remarkable backdrop to modern life. Images and accounts of poverty surround us through abundant fund-raising campaigns and other media. We are lured to think that we *know* poverty; it has lost its shock value. While the child mortality rate in the DRC – with 20 per cent of children dying before the age of five (Requejo 2010: 76) – is quite spectacular and presents a desolate reflection of just how well founded the fears and worries of the rape survivor are, even child mortality remains, it seems, too ordinary and familiar to attract our sustained attention in the DRC context.

Hence, the failure to listen is also a manifestation of a desire for an Other in need of being saved by 'the fitter Self' (Spivak 2004). Here, the image of the raped women fulfils special needs; it provides a particularly rewarding object of rescue for 'the fitter self' in that, in contrast to combating poverty, it imbues the fitter self with an air of particular heroism deriving from the supposedly enormous courage required to enter the violent, barbaric DRC war zone to meet these women.[18] The media attention and the heavy celebrity involvement also provide 'the fitter self' with an air of glamour and a chance to feel like and even be a celebrity (possibly paving the way to Oprah's sofa and the *New York Times*, as was the case with Lisa Shannon). In short, the partial listening and engagement in sexual violence is accompanied by several benefits that would be lost with a more comprehensive listening and engagement.

Let us now venture into the more concrete, and turn our attention to how this engagement was translated in actual interventions designed to alleviate the plights of rape survivors. How have these been designed and with what consequences?

The commercialization of rape

The first international report on sexual violence in the DRC was published in 2002 (Human Rights Watch 2002). A couple of years later, sexual violence emerged as *the* issue to be engaged in for a range of actors, from politicians, journalists and donors to international and national NGOs and researchers. In the words of Žižek (2009: 9), the 'SOS call' came that 'we have to act now!', 'drowning out all other approaches', signalling that 'everything else can and has to wait'. The raped women of the Congo emerged as the emergency within the emergency. They were singled out as the precious bodies to be mended and saved before others in the exercise of power over life (cf. Agamben 1998; Duffield 2007). Stories of emergencies tend to identify *the* favoured lives, *the* prioritized bodies whose lives are construed as the most precious (Duffield 2007). In Liberia, it was the child soldiers (see Utas 2011). In Sierra Leone it was the amputees (see Hoffman 2011). In the DRC it was the raped women.

Partial intervention As Autesserre argues, the Rape Story provided the quintessential simple and seductive narrative needed for successful fund-raising and advocacy in the DRC (Autesserre 2012; see also Chapter 2). The global attention to rape soon became a lucrative source of attention, goodwill and resources for donors and international NGOs who flocked to the eastern DRC in order to establish programmes for rape survivors. Reporting on and engaging in the issue of sexual violence became a means of accessing donor funding for NGOs, both foreign and national (Eriksson Baaz and Stern 2010). Following in the wake of increased donor funding, there was an exponential upsurge of local and national NGOs attending to the issue of sexual violence. While these include NGOs that previously worked and continue to work also in other areas, the number of single-issue NGOs with the aim of combating sexual violence rapidly multiplied – to, in the words of Douma and Hilhorst, 'pitch in the sexual violence market' (2012: 33).[19]

Sexual violence became 'the buzzword' to insert in project proposals in efforts to increase chances of securing funding (Autesserre 2012: 13). Given the attention and goodwill that engagement with sexual violence generated, the intervention field seems to have been characterized by a particularly fierce competition between intervening actors, who all struggled to demonstrate their own commitments and achievements. This competition was also evident between UN agencies themselves, which competed to be the prime intervening agency in relation to sexual violence. This has been reflected in a reluctance to share information and also (sometimes systematically) withholding information from the coordinating body.[20]

Moreover, organizations often resorted to 'victim appropriation': they treat women as *their* victims and 'try to hide the identity of beneficiaries out of fear that these will be poached by other organisations' (Douma and Hilhorst 2012: 28). As Douma and Hilhorst conclude, this raises 'a serious ethical question on treating victims as "commodities" in terms of project-funding. Several respondents used words like "force" and "kidnap" to describe how some organisations take women from their villages to put them in urban-based shelters to give the organisation the physical possibility to showcase them to donors' (ibid.).

The particular attention to sexual violence implied that, comparatively (with other intervention areas), massive resources have been channelled by international donors into sexual violence – particularly providing services to rape survivors (Eriksson Baaz and Stern 2010; Autesserre 2012; Douma and Hilhorst 2012).[21] Our discussions with international and national NGOs working within other areas of intervention bore witness not only to a lack of interest, but to difficulties in receiving funds.[22]

In this process, sexual violence was addressed in isolation from various related areas, such as women's general sexual and reproductive rights. For instance, rather than funding comprehensive sexual and reproductive health services, monies to hospitals have often been earmarked specifically for rape victims (Eriksson Baaz and Stern 2010; Douma and Hilhorst 2012). Moreover, sexual violence has largely been separated from its context of women's general legal rights, in that funding to organizations providing legal support to women has been earmarked specifically for rape victims, and not for all women who consult legal clinics (ibid.). Separate mobile courts aiming to end impunity in relation to rape committed by men in uniforms were established, ignoring other violence committed by armed and non-armed actors (to be further discussed below).

While many national and local NGOs (as INGOs) have contributed to the singular focus on sexual violence, several representatives of national and local organizations addressing women's rights issues whom we interviewed also criticized the one-sided focus on sexual violence (ibid.). In addition to frequently criticizing the inefficiency of outside interventions against sexual violence, not least the massive amounts spent (wasted) on facilitating international visitors and experts,[23] they also often described the one-sided focus on sexual violence itself as donor driven. A common assertion in these discussions was that sexual violence is merely one aspect of violence against women in the DRC, but that donors seem uninterested in other forms of violence. One representative of an organization providing legal assistance to survivors of sexual violence explained as follows:

> There are so many different aspects of it [violence against women]. There is domestic violence, that is a very big problem that we have not yet started to address. Women's rights are violated on a massive scale in this country. Also a huge problem if you talk to women in the villages is all the problems connected to inheritance and property rights. [...] They [donors] are mostly interested just in sexual violence, especially the ways in which it is used as a weapon of war and all that. And it is. But sexual violence is also committed by civilians, and in our very houses. But it is difficult to get money for other projects. So there is a certain difference in views between us and the donors. But if you discuss too much, and try to get in other things that they don't think are important, you might miss the funding [laughing]. So you avoid discussing too much with them. (Cited in Eriksson Baaz and Stern 2010: 55)

Many national and local women's organizations we consulted identified women's political participation and women's economic power, particularly inheritance and property rights, as the issues of utmost importance.[24] Several organizations pointed out that a majority of women who come to NGO legal clinics seek support for problems related to their inheritance and property rights. However, as they explained, given that donor funding was earmarked for rape victims, they were often unable to assist these women. Yet, as some representatives contended, they tried to find ways to support these other women as well with these earmarked funds. We will return to this below.

In short, an industry was created around the issue of conflict-related rape in the DRC. This process also, unsurprisingly, as we will discuss below, came to shape how ordinary people in the eastern DRC navigate within this industry in their search for survival.

The everyday commercialization of rape Mats Utas has argued that women (as well as men) in the war zone can be understood as 'tactical agents engaged in the difficult task of social navigation' (Utas 2005: 426). Through perceiving people in war zones as 'social navigators' we can better understand 'their search for protection of self and families, and the role "victimcy" can play in achieving these objectives' (ibid.). In the DRC, a singular focus on sexual violence has largely created the framework in which certain forms of social navigation have been enacted. More specifically: a singular focus on sexual violence has shaped a context in which rape (or allegations of rape) becomes increasingly entangled in survival strategies, and in which women are encouraged to represent themselves as survivors of rape in order to establish themselves as legitimate recipients of humanitarian aid.

The strategy of victimcy in relation to rape operated in various areas. One related to rape-induced injuries that require surgery, such as fistulas. Contrary to what perhaps could be concluded from articles and reports on rape and fistula, only a tiny proportion – 0.8 per cent – of fistula cases in the DRC are 'rape related' (Onsrud et al. 2008: 265). The vast majority of cases are caused during traumatic delivery. However, as noted above, donors have often earmarked healthcare, such as fistula surgery, for women classified as survivors of sexual violence. Consequently, in many hospitals only raped women were eligible for free treatment, while others had to pay sums they simply could not afford – compelling women with fistulas related to childbirth to present themselves as rape survivors to get access to surgery and other medical treatment.[25]

The impetus to 'invent' or embroider stories of rape in order to gain access to services has not been limited to the realm of healthcare. Similar scenarios played out in connection with other services provided for rape survivors, such as food aid, education programmes and credit facilities. According to local organizations consulted, such instances mostly involved women and girls who had lost or become estranged from their families and therefore had little to lose from the stigma of rape.[26] However, with the increasing attention to sexual violence and efforts to combat disgrace and shame, in combination with the high prevalence of such violence, the stigma associated with rape has also decreased (thus facilitating the process of presenting oneself as a rape survivor in order to access services). Moreover, since it became common knowledge in many areas that claiming rape-victim status renders you a legitimate recipient in the eyes of intervening actors, such a claim vis-à-vis donors does not necessarily mean that one is perceived and/or stigmatized as a rape victim in one's community (Eriksson Baaz and Stern 2010; Douma and Hilhorst 2012).

It should be pointed out that the story of the raped women in the DRC, while spectacular (and lucrative), is in no way unique in these respects (save perhaps for its allotted status at the top of the UN security agenda). As noted above, most war zones have their 'emergencies within the emergency'. A quite similar process of rapid horror-fame-profit-spectacle-masquerade occurred, for instance, in relation to the amputees in Sierra Leone. While a majority of the victims of RUF mutilations died, a few thousands survived, and they ended up in camps in Freetown. In a similar manner to the two main hospitals in the eastern Congo that attend to rape survivors, these camps became the target of 'an endless parade of people trying to capitalize on the publicity value of the amputees' (Médecins Sans Frontières, cited in Polman 2011: 62).[27] Hoffman (2011: 169) describes this process:

> The largest of these, the Aberdeen Road camp, became a required stop for visiting journalists, NGO workers, and dignitaries. No single space became more synonymous with the horrors of the war in Sierra Leone. Consequently, no space received more assistance than this camp. Relief organizations and charitable donors targeted the camp for funding projects or brought cash-donations, and it soon became clear that being one of the war's most tragic victims had ironically become one of the few guaranteed means of generating an income. The population of the camp swelled as destitute family members showed up in the capital to live with their mutilated relations. Despite the fact that a number of organizations not only donated artificial limbs to residents of the camp, but also showed them how to cheaply and effectively manufacture their own, almost no one in the camp chose to use them.

Hence, being selected as the 'precious life to be saved before others' propelled the amputees to develop 'a mode of narrating and displaying their violations prominently' in order to sustain themselves and their families (ibid.).

Let us now return to the DRC and the funding for combating and redressing rape and its manifestations. In addition to creating a situation in which some women have felt forced to present themselves as rape victims in order to access basic services, the singular attention to and funding for working against and remedying sexual violence contributed to a process in which allegations of rape became a particularly effective extortion and bargaining strategy. The problem of allegations of rape being used as a bargaining chip was raised in interviews with several people, both military personnel (see Eriksson Baaz and Stern 2010) and local organizations. One manifestation of this is (mostly male) family members falsely accusing men (often their daughters' boyfriends) of raping their daughters in order to extort money.

This phenomenon has increased with the many mobile courts that have toured the countryside of eastern DRC, funded by donors eager to get quick results on impunity in relation to rape committed by men in uniform (but not committed or patient enough to do this by a more long-term strengthening of the ordinary justice system). Most of these courts have exclusively addressed sexual violence, and the civil damages to be paid by convicted soldiers/combatants ordered by the courts are quite substantial. While these sums often have only a symbolic value since only very few of those convicted could ever pay even a fraction of these damages, the trial and threats of trials opened up new opportunities for adults to claim damages from men who had sexual relations with their daughters. Further, the revised law of 20 July 2006[28] provides more opportunities to do so, in that it stipulates that sex with anyone under the age of eighteen is considered to be rape.[29] As a result, many mobile courts heard cases involving a girl just under the age of eighteen, where the defendant was her soldier boyfriend with whom she had had consensual sex and the girl's parents were the accusers.[30] However, there are also accounts of women who have sexual relations with men in order to be able to accuse them of rape, sometimes in collaboration with the police (Douma and Hilhorst 2012).

In sum, the focus on sexual violence as a *particularly serious* crime enables people from various groups – from the police and justice personnel to civilians – to use allegations of rape as an income-earning strategy. The 'SOS call' to 'act now!', 'drowning out all other approaches', signalling that 'everything else can and has to wait' (Žižek 2009: 9), led to troubling manifestations in the DRC (encouraging women to present as rape survivors in order to access basic services and false rape accusations as an income-earning strategy). While this is certainly troubling, even perverse, it is nonetheless unclear that the destructive ramifications of a singular and well-funded focus on interventions aimed at redressing sexual violence call for a cessation or even a redirection of these interventions. Our unease as participants in and witnesses to the commercialization of rape (with its rapid horror-fame-profit-spectacle-masquerade dynamic) have nonetheless induced us to wander in the ethical morass that this dynamic presents. What are the (potential) consequences of highlighting these perverse implications of the global attention paid to sexual violence? In whose interest do we – as 'critical scholars' – speak here? Does it actually benefit Generose and her 'raped sisters' (whose interests we somehow claim to represent)? These are some of the thorny issues to which we will now attend.

Who speaks, and who is complicit (and in what)?

Let us start with the issue of the commercialization of rape (the ways in which we highlight and critique a situation in which some women feel forced to present themselves as rape victims in order to gain access to basic services and assistance). Do we not risk reproducing the classic image of African aid partners and beneficiaries as unreliable cheaters, which is already so prevalent in development discourses and so clearly linked to the colonial library (Eriksson Baaz 2005; McEwan 2009)? Do we not risk simply confirming the already highly problematic (and often racist) images some donors hold of the 'locals' – donors who often display a very low capacity for self-criticism to begin with and instead abrogate their own responsibility and put the blame for 'failure' on what they deem as the passive, unreliable, non-committed, cheating 'partners' and beneficiary Others?

After all, the unwelcome question that our critique elicits reads as follows: are we not simply saying that some Congolese women sometimes lie about being rape victims? Unfortunately, our reluctant 'yes ... well, sort of, but not really ... and that is not really the point', which we utter in the hope of re-directing the problematically framed question, risks translating into a firm 'yes', therewith construing Congolese women as unworthy of further funding – for at least some within the (diverse) donor community. Indeed, might the critical remarks on the problematic manifestations of the singular focus on sexual violence actually contribute to further impoverishing Congolese women by interjecting a reduction of funds (through the logic: 'if it is not really that much of a problem, and if women are actually even lying about this, why should we be involved in this at all')? This warning about the danger of our critique found favour with the former UN Special Representative on sexual violence in conflict, Margot Wallström, who has persistently denied the problem of commercialization and firmly argued that the singular focus on sexual violence is not at all problematic, but attracts attention also to other issues identified by Congolese women.[31] If the former UN Special Representative is right, it means that such a critique risks contributing to less attention – and potentially less funding – for these other issues identified by many Congolese women's organizations.

So far, and unfortunately, nothing indicates that this position is warranted. Although it is of course difficult to assert with any certainty, the focus on sexual violence seems not to have resulted in increasing funds to other areas. Even in 2011 (when the attention given to sexual violence could be expected to have been going on long enough for potential 'trickling down effects') Congolese and foreign aid workers 'regularly complained that they cannot draw the attention of the media or donors to horrific events that have no sexual dimension' and 'that they receive more money than they need to treat victims of sexual abuse, while they lack funding to implement other crucial projects' (Autesserre 2012: 15).

However, even if the focus on sexual violence does not attract attention and funding for other issues also identified by Congolese women's organizations, these organizations, as well as Congolese women themselves, are clearly not passive recipients of interventions. Instead, they are active agents who strategize in their dealings with various intervening actors. As reflected in the quotation above from a local organization, this resistance rarely takes the form of an overt collective undertaking, which can be found in many similar local contexts. Rather, it manifests itself as a veiled, unarticulated resistance through efforts to create a space for manoeuvre within the rules set up by the donor (Crewe and Harrison 1998; Eriksson Baaz 2005; see also Hansson 2012 and Hansson et al. forthcoming). One manifestation of this is using the issue of sexual violence to secure funding and then – in various ways – making use of the earmarked funding so that it can also benefit other women.

Consequently, Margot Wallström might indeed be mistaken in claiming that the focus on sexual violence attracts attention and funding to other issues. Nonetheless, Congolese women's organizations (such as some INGOs), by force of their own agency, sometimes managed to use the funds they acquired through the 'sexual violence' channel to also benefit other vital areas which would otherwise not have received support. In light of this, our critique surely does risk contributing to a situation of decreased attention and funding – for all areas. After all, it is the Rape Story that enabled the DRC conflict to receive some global attention and consequent donor support to mitigate its violent effects. As Autesserre (2012: 10) concludes in her recent study, since 2009 'there has been no interest in the Congo at the UN Security Council except when it discussed incidents of mass rapes and potential responses to them'. Similarly, 'US State Department top officials reportedly pay no attention to the Congo except when sexual violence grabs the headlines'.

Surely, any responsible critique must take into account the fact that 'African security issues […] struggle to register on western news organizations' agendas' (Michalski and Gow 2007: 167), that 'Africa does not count' – or at least, 'the exchange rate at which it does come to count is enormously high and painful in terms of human costs' (ibid.). As Frank Humphreys shows, the DRC conflict (dominated by the Rape Story) has, in comparison with other stories in the global media landscape, received only a tiny fraction of attention. According to figures he presents from the global media covered by the Alertnet world press tracker, between September 2006 and April 2007, only 1,327 stories focused on the DRC, while the Israel–Palestinian, Afghanistan and Iraq conflicts generated 19,946, 29,987 and 43,589 stories respectively. While the peak of the DRC Rape Story came after 2007, this difference in media coverage in relation to the cost of human suffering is still astounding. In light of this, his query as to whether 'sensationalism (whether accurate or not)' is 'the alternative to "The Greatest Silence"?' is indeed pertinent (Humphreys 2011).

In this context, is our critical reading responsible? Would not several local organizations, as well as many Congolese women themselves, prefer sensationalism to a critical appraisal that risks contributing to silence? Would they not prefer and appreciate people like Lisa Shannon and Eve Ensler, who, despite the above problematics, do bring in some desperately needed funds that can benefit not only rape survivors but also others (through the agency of local organizations as well as women themselves) through their well-meaning engagement. In short, is it not wiser to accept the sensationalist and sometimes racist workings of global media and humanitarian action (which, after all, brings attention and funding) and acknowledge and trust the capacity of national and local organizations and women (and men) to make (good) use of these funds – and simply refrain from criticism?

There is of course no easy or absolute answer to this question. However, as is evidenced in our decision to write this chapter, our answer would be 'no'; we deem that a healthy dose of criticism is warranted. First, we believe that rape is much too serious to be rendered banal through its commercialization and the attending questioning of the credibility of women who are survivors of rape. The one-sided interventions that force some women to present themselves as rape victims in order to access basic services, and which encourage false allegations of rape as a survival strategy, we argue, contribute to a banalization of the act of and suffering induced by rape itself. Such banalization is a process that can only be counterproductive for long-term struggles against sexual violence.

Furthermore, considering the silenced voices of those who are subject to both rape and its remedies, as well as the spin-off effects of the rapid horror-fame-profit-spectacle-masquerade dynamic, demands that we also query what the new-won attention to wartime sexual violence fails to deliver. We are compelled to interrogate the unequal relations of power which are concealed in its luminous rhetoric of global sisterhood.

Failing to dare to explore the confines and pitfalls of seemingly progressive feminist engagement is to take a truly paralysing position. Indeed, compliance with and complacency towards the commercialization of rape, which coincides with and is preconditioned on critically needed attention to the harms of sexual violence, would entail an implicit acceptance of racist and imperialist conceptions of 'the burden of the fittest' (Spivak 2004: 538). Furthermore, shying away from the (self-manifested) call to critique would mean contributing to reinscribing the dominant portrayal of the white woman as the hero and saviour of women in the Congo and an acceptance of the marginalization and silencing of the multiple voices of women (and men) in the DRC.

Nonetheless, our 'no' (to whether it would be more responsible to refrain from criticism) is not – and cannot be – an easy one. In our 'attempt[s] to look around the corner, to try to see ourselves as others would see us' (Spivak 1999: xii–xiii) the somehow uncomfortable thought that the rape survivor as well as the local organizations attending to her probably prefer Lisa Shannon or Eve Ensler to us has recurred constantly. In contrast to them, we certainly fail to bring much that is useful to ease the day-to-day plights of rape survivors.

We have been reminded of the tendency of many post-colonial critics, including ourselves, to engage in an oversimplified, self-righteous critique of others' faulty or hypocritical engagement – without contemplating the potential effects this criticism might have for people whose survival (at least to some extent) depends on such engagement.[32] We have been (uncomfortably) reminded by some women and local NGOs of the fact that they 'eat' thanks to – and thereby value (and also reproduce) – the very paternalist/maternal colonial representations and interventions that we criticize. As critics of humanitarian and development interventions, we must remind ourselves of our own message to the interveners; that good intentions (in this case, revealing and interrogating the myriad power relations and violences embedded in the attention to wartime sexual violence) 'are not enough'.[33]

We must also remind ourselves that we are ourselves deeply complicit in the processes and power relations that require our critical gaze. We have (particularly initially) certainly been complicit in contributing to and profiting from the singular attention to wartime rape in the DRC warscape. Engagement in wartime rape in the DRC has served as a source of attention, goodwill and resources not only for international donors, journalists, politicians, etc., but also for researchers, such as ourselves. While we have received a bounty of funding and attention, researchers working with less fashionable issues, such as child and maternal mortality in the DRC, have faced a very different situation.

In this chapter we have nonetheless endeavoured to emphasize otherwise silenced concerns (as we have heard them) over, for example, the endemic lack of basic means of survival and the workings of other forms of violence from which people suffer. We have done so since these concerns, which we have heard repeatedly since 2006, have been rendered mute in most stories of violence in the DRC. Our efforts to do so, however, are certainly flawed. As we noted in Chapter 1, we banged our heads against the limits of familiar imaginaries in our attempts to hear the rapists' stories. So too were we certainly limited by being blind and deaf to the stories and agency of rape survivors and Congolese women's organizations. Lisa Shannon is not alone in inviting the critique that she propagates a 'mood of self-congratulation as [a saviour] of marginality'. The same can certainly be argued for critics such as ourselves. We are also guilty of the failure to 'unlearn our privilege as our loss' and cannot claim authority to 'speak for/represent' the 'rapists', the 'raped woman'.[34] We have surely also failed to 'listen from below'.

5 | Concluding thoughts and unanswered questions

Throughout the different chapters of this book, we have sought to unravel the dominant storyline of sexual violence as a 'weapon of war' in order to explore its confines and open up space to think otherwise about conflict-related sexual violence and its subjects. In distinct yet complementary ways, we have reflected on its underlying assumptions and ontologies, studied its composition and exclusions, identified some of its lacunae and contradictions when seen through the prism of military sociology and the sociology of violence, as well as in light of the experiences of the DRC warscape, and interrogated the ethico-politics of both research and humanitarian engagement under its framing. When we reflect on the book as a whole, we are struck by how even the most confidently penned critique also prompts an afterthought of doubt, even trepidation. Why is a critical reading of the dominant, appealing – and surely progressive – framing of wartime sexual violence so intricate and difficult? And what have we learned by undertaking this endeavour? We answer these questions through a retrospective reflection on the main points (and puzzles) raised throughout the book.

Sex/gender and the creation of uncomfortable subjects

We felt compelled to begin our journey in Chapter 1 by returning to critically reflect on our own research process, whereby we focused on the 'perpetrator' instead of the victim in our attempts to better understand sexual violence among the national armed forces in the DRC – the supposed 'rape capital of the world'. Indeed, this starting point presented itself as necessary, as we could not let go of the puzzles and dilemmas that conducting such research posed. They demanded working through. This entailed interrogating the underlying assumptions and ontologies of the framing, which we struggled to peek beyond. Our efforts on this part of our journey yielded the following understandings.

The main storyline presents wartime sexual violence as a particularly heinous crime (as *sui generis*, of itself), and casts rape (and those who enact rape) as exceptional, and simultaneously both normal and abnormal – yet in slightly different registers. We thus identified a double move, whereby sense is seemingly imparted and a moral compass installed from and by which we can adjudicate and act (see also Stern and Zalewski 2009). First, the dominant storyline infers that rape/rapists are 'normal' through several interrelated registers. The 'Sexed' Story of rape haunts the accepted 'Gender' Story, and rape as a 'natural' by-product of male heterosexuality gone awry in war settings appears as naturalized. Racialized notions of rape as Other and barbaric enable rapists to be cast as 'normal' to an antiquated, natural, bestial state (yet simultaneously profoundly abnormal to civilized society). Additionally, in a different register, rape emerges as the fault of gender and its manifestations in the (normalized) workings of the military; gender materializes as knowable and 'fixable': normal. Furthermore, rape/rapists emerge through their very failure to be 'normal' – which is arguably a 'normal' predicament inherent in the fulfilling of any ideal subject position.

In the casting of rape/rapists as abnormal (and incomprehensible), those who rape are stripped of their humanity. This is hardly surprising. What we have argued, however, is that it is *through* the workings of gender (seemingly opposed to, yet nevertheless haunted by, 'sex') that such dehumanization of 'perpetrators', 'victims' and ultimately even rape as a human activity occurs. Furthermore, and importantly, by the singular focus on rape as a violent act *sui generis*, other forms of violence are normalized, even naturalized. This double move – the casting of rape and rapists as both normal and abnormal – with its attendant ethical shorthand reveals a host of uncomfortable subjects who are difficult (and uncomfortable) to see and hear – such as the rapist/beast/non-human self. It therewith thwarts certain types of research, listening and empathy. We therefore also touched on the persistent conundrum of complicity in researching violence and those who commit violence given the above.

The lure of a single route to redemption

After laying the groundwork in Chapter 1 through our exploration of the relations between sex, gender and violence, in Chapter 2 we turned to examining the composition of the Rape as a Weapon of War discourse. Building on this groundwork, and through a wide and in-depth reading of texts that reproduce this discourse, we were able to better interrogate both its appeal and its purchase. As we have noted many times throughout this book, the Rape as a Weapon of War discourse breaks with the view of rape as a tragic but natural and inevitable outcome of war unleashed by men's inherent sexuality, and instead casts rape as avoidable. Simply put, the Rape as a Weapon of War discourse promises deliverance and even retribution. Through its actualizations of what we have called 'strategicness', it sharply diverges from previous narratives of wartime rape (the 'Sexed' Story) and its gloomy prospects for change, which are incompatible with current official (politically correct) ethics and norms of warfare.[1] By contrast, the discourse of Rape as a Weapon of War assures us that change is not only possible, but imminent.

How, in short, does this work? Rapists are (re)cast as rational modern subjects who can be held accountable for their rational choices to implement the strategy and wield the weapon of rape. Through actualizing 'strategicness', clear ethical distinctions are rendered and responsibility can be borne. Indeed, casting sexual violence as strategic, systematic, rational situates the gendered violent subject in a moral world whose contours we recognize. His weapon can be mastered and regulated as other weapons can. The Rape as a Weapon of War discourse thus assures us that we need not be out of control and that rape and rapists can be disciplined, and misogynist societies can evolve. The Rape as a Weapon of War discourse thus promises change and deliverance along a sure teleological trajectory, which meets the urgency that our newly acquired awareness of rape as 'planned and orchestrated as a tactic of war' (Wallström 2010b) entails. Hence, its widespread status and recognition as not only the most accurate but also the most ethical narrative cannot be understood outside its assurances regarding change and redemption. Yet our critical reading also showed us how, despite its teleological seduction and political importance, this discourse remains unstable and its promise of deliverance, precarious.

Leaking military structures and the uncertainty of war

To further probe its fragile yet universalizing composition, in Chapter 3 we read the Rape as a Weapon of War discourse through the lens of military sociology, theories of violence and an in-depth analysis of the case of the DRC. In particular, we sought to interrogate the ways in which strategicness was actualized in terms of the presumed general intentionality and functionality of wartime rape – whether on the level of strategy, tactic or weapon. In particular, we highlighted the discursive nature of strategicness by discussing how military notions of strategicness vary, and how, in some contexts, military commanders do not simply encourage but rather try to prevent sexual violence – since they construe it as non-strategic (see also Wood 2009, 2010). However, as we concluded, notions of whether rape is strategic or not cannot be framed as reflecting a simplistic division between greed and grievances.

In addition, we addressed the workings of military institutions and underscored how, in contrast to the dominant representation in the Weapon of War discourse, military institutions rarely function according to the celebrated ideals of discipline, hierarchy and control. Reflecting both the agency of soldiers themselves and the defining uncertainties of warring, the efficient, rational military remains an ideal that has turned out to be difficult to attain in practice, particularly on the battlefield. One of the predicaments of the discourse of Rape as a Weapon of War (as well as feminist research on militarism more generally) is that it cannot encompass such 'failures'. Instead, out of necessity in terms of the cohesion of its narrative, it tends to portray (and reproduce) the military institution as the rational war machine such institutions aspire to be.

Additionally, building on Kalyvas (2006), we explored how violence against civilians, including sexual violence, may stem not only from military tactics and the meta-narratives of war, but also from the micro-dynamics of violence in war. Indeed, our reading leads us to call for an acknowledgement of how violence in war, including sexual violence, is also related to micro-level dynamics and private dispute settlement, involving both armed and non-armed actors. Moreover, those who seek to understand wartime rape would do well to also recognize the ways in which violence (including sexual violence) is shaped by the situational and interactive nature of warfare and the workings of forward panic (Collins 2008) and spirals of violence (Muñoz-Rojas and Frésard 2004; Staub 1989).

Our foray into military sociology and the sociology of violence, as well as the specific context of the national armed forces in the DRC, draws attention to how the discourse of Rape as a Weapon of War ascribes too much rationality to acts of sexual violence. As Kalyvas (2006: 32, 33) notes, accounts of violence tend to 'cluster around two poles': one 'descriptive, rich and highly dramatic, associated with a view of violence as an irrational and atavistic pathology, whereas the other takes violence to be an outcome of narrowly instrumental goals, with a tautological bent'. In the latter, '[m]ad subjects are replaced by instrumental leaders able to manipulate myopic citizens and implement policies of violence to achieve their goals'. Building on our analysis in both Chapters 2 and 3, we argue that the discourse of Rape as a Weapon of War is situated at the 'tautological pole'. However, by highlighting the complex and messy nature of violence in war zones, as we have done here, we by no means promote a 'clustering' around the 'Hobbesian pole', which represents violence as irrational and archaic. We simply point to the tendency to attribute too much rationality to sexual violence in war and the need to avoid a simplistic, tautological line of argument whereby we assume that rape occurs simply because it is useful to military and political leadership. Indeed, the main purpose has been to call for attention to complexity and context in understanding and ultimately redressing conflict-related rape. Reductionist and generalized conceptualizations of Rape as a Weapon of War are problematic, not only because they overshadow other forms of violence, but because they effectively conceal other factors that contribute to conflict and post-conflict sexual violence. The Rape as a Weapon of War discourse certainly does offer much valuable insight into some of the dynamics of warring (and gender) in particular contexts. Nonetheless, the way it reduces sexual violence simply to a 'weapon of war' limits the possibilities of understanding the various factors that may converge to contribute to, or be conducive to, such violence.

Turning back the clock?

However, troubling questions surely arise from our problematization of the discourse of Rape as a Weapon of War in this book in this way. First, does a critique such as ours not invite a regression into the paralysing and complacent state of unavoidability, familiar to that presented in the 'Sexed' Story? Does not rape, yet again, become a regrettable side effect of war? We respond to this question with a firm 'no'. First, such a rendering accepts a simple teleological trajectory of progress, which neglects the ways in which such teleologies depend on the reconstitution of the previous state from which one develops in the very writing of the temporal notion of development and possible regression. There is no 'there' to which to 'return'. Furthermore, as Wood (2009, 2010) demonstrated, rape is committed to varying degrees in different conflicts and is nearly absent in some. Hence, empirical evidence itself already suggests that rape in war is avoidable; writing rape in war as inherently strategic is not the only possible route to redemption.

Another disturbing question arising from our critical reflections concerns the responsibility of perpetrators: in situations where commanders do not control their troops, and where violent acts are committed as a result of forward panic, are commanders/perpetrators really responsible? Do we not suggest that perpetrators often cannot be held responsible and that they therefore should not be convicted for their crimes? Again, our response is a firm 'no'. Agency and attendant ethical responsibility reside in the ways in which we act (and how those acts affect others) when navigating and mediating different governing discourses (Braidotti 2006: 14; Campbell and Shapiro 1999; Jabri 1998, 2004). Furthermore, the juridical answers to these questions are certainly not best answered by us.[2] As Paul Higate explains, in contexts 'where perpetrators are responding to situational and interactional forces within the context of a well-honed proficiency in violence unleashed inappropriately and with devastating consequences through a changed state in the tunnel of forward panic [...] the perpetrators might not be seen as the autonomous decision making agents constructed in law' (2012a: 26). However, this does not end the 'call for responsibility' (ibid.). Such responsibility surely resides with those who enact violence and inflict harm, although the parameters of such responsibility do not only include them.

What we 'cannot not want'

In order to further query questions of responsibility (but from a different angle), in Chapter 4 we turned our gaze towards the victims or survivors of sexual violence and to our efforts to attend to their pain and their needs. A post-colonial reading of the global battle to alleviate the suffering of the raped women in the DRC revealed a host of troublesome effects of engagement. The voices of women who have been raped have been often ignored by interveners, whose interest, for the most part, is directed towards one aspect of their story – that of rape. Other parts of their stories – their lack of basic means of survival; concerns over lack of land to cultivate, lack of clean water, concerns over hungry and often malnourished, sickly children – simply often remain unregistered.

Moreover, our and others' research has also indicated that wartime rape has become commercialized; donors as well as international and national NGOs have capitalized on the global attention accorded to rape in order to receive attention, goodwill and resources. Moreover, the subsequent singular focus on, and funding for, sexual violence has been manifested in an everyday commercialization of rape in the DRC. Allegations of rape became deeply entangled in livelihood strategies. People feel compelled to claim 'victimcy' to be eligible for humanitarian funds; false accusations become an effective income earning strategy.

However, highlighting such perverse workings of the engagement in wartime rape is clearly not simple, but paired with several dilemmas. One such dilemma is related to the issue of our (and others') responsibility in how the phenomenon of wartime rape is framed, represented and critiqued. Is it responsible, we have often wondered and been asked, to criticize the manifestations of a great victory for feminist progressive politics; to critically interrogate the workings of what has clearly been a monumental and vitally important accomplishment: the global attention given to wartime rape and its victims? Do we not risk contributing to a renewed silence? Would not keeping quiet be more responsible?

Another related dilemma involves the implications of our (and others') criticism of the workings of the 'rape industry' for women in the DRC. As argued in Chapter 4, such a critique surely risks contributing to less attention, and through this less funding, not just for rape survivors, but also for other issues related to women's rights. After all, it was mainly the 'rape story' that brought the DRC conflict its 'moment of fame' in global media and the political arena (and subsequently attracted funding). Moreover, many national and local actors have been able to navigate quite well around donor regulations and have managed to use earmarked funds for other areas identified as crucial for them, such as women's inheritance and property rights. In this context, it is indeed pertinent to ask whether our critical inquiry is ultimately harmful.

We respond to these (self-)critical questions with the help of Spivak. As she argued, critical analysis of equally deeply cherished concepts and victories is sorely needed. In her words (in relation to liberalism and other modernist emancipatory constructions), we must engage in 'a persistent critique of what one cannot not want' (interview in Landry and MacLean 1996: 28; Spivak 1993: 284). What we 'cannot not want' here is the newly won attention to wartime sexual violence. As we discussed in Chapter 4, we believe that wartime rape is much too important to be banalized (which is a consequence of the commercialization of rape in the DRC warscape). Hence our critical inquiry into the failures and tenacious manifestations of the attention to wartime sexual violence does not entail a questioning of the value of this attention. Rather, it signals a firm commitment to working to stop wartime sexual violence against women and men. However, it is also our responsibility (as well as that of other critical feminists committed to combating wartime sexual violence), we deem, to query what this newly won attention to wartime sexual violence fails to deliver. This includes examining the power relations that are veiled in even the most well-meaning feminist engagement.

What we also 'cannot not want' is the engagement *with* (not *for*) the women and men who have suffered from wartime sexual violence, in the DRC, as well as elsewhere. To be clear, we are not arguing that the survivors of rape in the DRC should have been isolated from the outside world. Nor do we propose that Lisa Shannon, whom we introduced in Chapter 4, should have stayed in the USA to tend to her photo business instead of working for rape victims in the DRC. Our argument is simply that engagement can – and indeed must – be different. Our critical enquiry is positioned in the necessity to commit to 'building a noncolonizing feminist solidarity across borders' (Mohanty 2003: 224). Given that 'colonizing feminist interventions' rely upon deeply cherished privileges and grant the rewards that the position of 'saviors of marginality' bring, this is clearly a taxing task. Failures of 'learning to learn from below' are surely not simply manifestations of the inherent difficulty of this task. Instead, such failures surely reflect a 'lack of will' – particularly in terms of paying the price that such a commitment entails.

However, in spite of these immense and daunting challenges – and regardless of how utopian a 'noncolonizing feminist solidarity across borders' appears from the horizon of the DRC – these challenges can surely not be used as excuses for not embracing such a vision. We can learn to better 'learn to learn from below'. Our listening can be less partial. We can learn to become more open to the twists and turns in the stories we hear, to be open to the 'unexpected response' (Spivak 2004: 537) and imagine 'the possibility of being somewhere that is not the Self' (Spivak 1997, cited in Kapoor 2004: 642).[3] We can stretch ourselves to think *otherwise*, as we explored in Chapter 1. Moreover, we can learn to better resist the temptation to produce the 'subaltern' to suit our own interests, to resist occupying the centre stage and writing ourselves as indispensable saviours. We can do better in reconfiguring our understanding of responsibility from 'the duty of the fitter self "toward less fortunate others" – to a responsibility *to* the other'; 'the predication of being-human as being called by the other, before will' (Spivak 2004: 535). Commitment to this vision, we believe, can be achieved only through a continuous critical reflection on our repeated failures to do so. Here, we (the authors of this book) have surely also failed. Yet, we suggest, in the recognition of failure lies the hope of ethics, of the political, of humanity.

Notes

Introduction

1 Most notably, Security Council Resolutions 1820 (UNSC 2008) and 1888 (UNSC 2009a).

2 We are not alone in voicing concern over the dominant framing of rape as a weapon of war. For instance, see also Buss (2009); Kirby (2012); Wood (2009, 2012), among others.

3 For a discussion of 'grids of intelligibility', see Butler (2004a); Dillon (2004); Foucault (2005).

4 We discuss our understanding of sex and gender in more depth in the following chapter, under the heading 'Sex/gender?' Suffice it to say here that in using the terms 'Sexed' and 'Gendered' as the titles for distinct narratives, we are reflecting the commonplace distinction between masculinity and femininity in terms of 'gender' (as constructed) in contrast to the notion of the biological differences of 'sex' as naturally given and (partially) determinate of masculinity and femininity. As we discuss later, we see such a distinction between 'sex' and 'gender' as highly problematic; any notion of 'sex' is produced within the social world. See Butler (1990, 2004b; Benhabib et al. (1995); Chambers and Carver (2008); Scott (1999); Riley (1988); Stern and Zalewski (2009).

5 We use the term 'strategicness' instead of strategy as an overriding notion that encompasses rape as a strategy, tactic, tool and function, which is employed for particular purposes in order to fulfil certain military aims. See Chapter 2 for a discussion of the terms.

6 See Nordstrom (2004) for a discussion of warscapes. See also Eriksson Baaz and Stern (2008, 2009, 2010, 2011, 2013).

7 The UN's Special Representative of the Secretary-General on Sexual Violence in Conflict, Margot Wallström, has described the DRC as the 'rape capital of the world'. See Wallström (2011); BBC (2010).

8 Eastern Congo has also become infamous as the 'worst place on earth to be a woman' (Viner 2011).

9 For instance, according to MONUC, over 1,700 civilians were killed in North and South Kivu during the military operations during 2009 and 6,000 homes were burned. Moreover, the issue of child soldiers has received comparably little attention in the DRC, despite the fact that 8,000 children are estimated to still be in the ranks of armed groups. For recent reports on the levels of 'non-sexual' violence, such as mass killings, systematic torture, forced recruitment, forced labour and arbitrary arrests, see, for example, Bureau of Democracy, Human Rights and Labor (2009); Human Rights Watch (2009c); Sawyer and Van Woudenberg (2009).

10 This project was financed by the Swedish International Development Cooperation Agency (Sida). It is based on interviews with more than 230 soldiers and officers between 2006 and 2011. Two-thirds of those interviewed (approximately 80 per cent men and 20 per cent women) were former government soldiers (ex-FAZ and FAC). The rest were integrated into the armed forces after the peace accord in 2002 and the first round of military integration, including Rassemblement Congolais pour la Démocratie (RCD), Mouvement pour la Libération du Congo (MLC) and the Mai-Mai. A majority experienced combat, either within the national army or within the

armed groups to which they previously belonged. While we did conduct some individual interviews, most interviews were organized as semi-structured group interviews (groups of three to four persons) with soldiers/officers from the same unit with the same rank and gender. This set-up (homogeneous groups, and group interviews rather than individual interviews) turned out to be the most fruitful, since it rendered the interview session less intimidating for the participants. Moreover, the group interviews – following the structure of the army itself, with people from the same rank who also knew each other – were advantageous in that the interview sessions often turned into open discussions and debates within the group itself. Maria Eriksson Baaz conducted the majority of the interviews in the national language Lingala (without an interpreter). Importantly, we view the texts that emerged out of the interview context as narratives co-constructed in a particular setting, not as accurate reflections of how those interviewed 'really are' or of their 'true' reasons for joining the military or participating in violent acts. For more details on this research project, see Eriksson Baaz and Stern (2009, 2010).

11 This research project was also funded by Sida.

12 Such a 'camp structure' (Sylvester 2007) is, however, starting to loosen. Indeed, this is already perhaps something of an outdated critique, as cross-references are beginning to figure in recent writing on sexual violence. See, for example, Kirby (2012); Skjelsbaek (2012); Wood (2012).

13 For some exceptions, see, for example, the work of Elisabeth Wood (2009, 2010), Paul Higate (2012a, 2012b) and Victoria Basham (2009).

14 Sense depends upon a temporary stabilization of meaning. Discourses offer a view or story of reality, which appears as natural, objective and taken for granted, and which effectively limits our understanding of the world. They are made up of signs, which are imbued with meaning through their relationship to other signs in the available repertoires of meaning. As Doty has explained, 'meaning is produced and attached to certain social subjects and objects, thus creating certain interpretive dispositions that create certain possibilities and preclude others' (Doty 1996: 4). Representations emerge as the meaning produced through associations (strings of signs and signifiers) within discourse. 'Representations that are put forward time and again become a set of statements and practices through which certain language becomes institutionalized and "normalized" over time' (Neumann 2008: 61).

15 It is the differences between concepts that signify; meaning is provided through deferral, through rejecting and excluding other possible meanings. These other possible meanings can be seen as necessary exclusions, or in Derrida's term the 'constitutive outside' that, as Mouffe explains, comprises 'the traces of exclusion which governs [...] constitution' (Mouffe 2000: 147).

16 Torfing explains as follows: '[...] the outside is not merely posing a threat to the inside, but is actually required for the definition of the inside. The inside is marked by a constitutive lack that the outside helps to fill' (Torfing 2004: 11).

17 Lacan's *point de capiton*, literally 'quilting points' (Lacan 1993). Nodal points (partly) fix the discourse through temporarily stabilizing the meaning of a chain of signifiers in order to impart sense (Laclau and Mouffe 1985: 112).

18 The fact that men are also raped surely complicates the notion that rape survivors are women. We address this 'uncomfortable subject' in Chapter 1.

1 Sex/gender violence

1 See also Doty (1993: 279–99), cited in Brassett and Bulley (2007: 10), for a further discussion on querying how violence is made possible.

2 There is, of course, a long tradition

of work that focuses on understanding perpetrators' motives and actions. See, for example, Brett and Specht (2004), Hatzfeld (2005), Horwood et al. (2007), Utas (2005), Staub (1989, 2011), Keen (2005), Kassimeris (2006), Fromm (1973), Bourke (1999, 2007), Groth and Birnbaum (1979) and Zimbardo (2008).

3 Our analysis will be complemented in Chapter 2 through close readings of several examples of the Rape as a Weapon of War discourse; here, we provide an analysis of the discourse in more general terms, in order to be able to flesh out the moves by which the sex-gender-violence equation is seemingly resolved.

4 On 'thinking otherwise', see Marysia Zalewski (2012), Robyn Wiegman (2003), Anthony Burke (2007: 30–1) and Jacques Derrida (1982, 1993).

5 We return to the critical question of the politics of listening in Chapter 4.

6 The predominance of this narrative in military contexts was also evident when conducting the DRC study. There were quite big differences in the ways in which civilians and military staff (both national and international) understood the problem of rape. While civilians (NGOs, INGOs, diplomats) tended to see it in terms of the Rape as a Weapon of War explanation, military staff, national and international, echoed the Biological Urge/Substitution Theory. Rape was explained as a result of opportunity, of boredom, sexual lust, the need for a distraction. For instance, in interviews both with US and EU staff, the most prevalent comment we heard regarding the frequent rapes committed by the FARDC was 'I tell them they cannot rape just because they are bored' (interviews with external, e.g. with representatives of the USA, the EU, the UN, as well as national actors in the DRC in October 2009, October/November 2010 and May 2011).

7 Hobbes's 'state of nature' is characterized as follows: 'no society; and which is worst of all, continual fear, and danger of violent death; and the life of man,

solitary, poor, nasty, brutish, and short'. Hobbes (1651: ch. XIII).

8 Parts of this section have been adapted from Eriksson Baaz and Stern (2009).

9 E.g. Stiglmayer (1994). See also Leatherman (2011), Pankhurst (2009), Skjelsbaek (2001), Kirby (2012) for overviews of how wartime rape is understood.

10 Portions of this section rely on a section with the same title in Stern and Zalewski (2009: 622), either verbatim or adapted for the purposes of our arguments here. We have not cited this work directly in order to render this text more readable.

11 There is no single universal definition of rape. Many definitions refer to the perpetrator, without specifying, yet implying that the perpetrator is male. For example, rape is defined in the appeals chamber judgment of the International Criminal Tribunal for the former Yugoslavia (ICTY) in the 2002 *Foca* case as '[t]he sexual penetration, however slight: (a) of the vagina or anus of the victim by the penis of the perpetrator or any other object used by the perpetrator; or (b) [of] the mouth of the victim by the penis of the perpetrator; where such sexual penetration occurs without the consent of the victim' (Human Rights Watch 2003: 2). See also Eriksson (2010) for an in-depth analysis of definitions.

12 See Butler (1990, 2004b); Benhabib et al. (1995); Scott (1999); Riley (1988).

13 See above note. See also Chambers and Carver: 'To understand sex as gendered is not equivalent to claiming that there is no difference between sex and gender. Making this last move amounts to collapsing the distinction between sex and gender so thoroughly that we have nothing left but gender. But while gender is lived through the body (as Butler constantly reminds us) gender norms are not *inherently* bodily. This means that to reduce everything to gender *is*, in a way, to do away with the body. One must hold on to a conceptualization of sex, so that

one does not lose sight/site of the body. In neither case does, however, this mean conceptualizing either sex or the body as natural. Nor does it imply that we think of sex as analytically distinct from or before gender. Sex is gendered. We only understand sex through the norm of gender' (Chambers and Carver 2008: 66).

14 See Lwambo (2011). See also Johnson et al. (2010).

15 See Horwood et al. (2007: 15) for an example.

16 For further discussion, see, for instance, Ali (2007); McClintock (1995); Sinha (1995).

17 As explained in Stern (2011: 34). Indeed, the colonial project can be seen as a civilizing mission, whereby Western civilization was presented as the universal end state in a modern teleological narrative; the familiar 'white man's burden'. Yet, importantly, in this grammar, the colonized must remain fundamentally different from the colonial identity. Bhabha explores this 'ironic compromise of mimicry', whereby the Self desires a recognizable and nameable Other who is 'almost the same, but not quite', and thus remains forever mired in inferior difference (Bhabha 1994: 86). This paradox also has a temporal dimension; Chakrabarty explains: 'the inhabitants of the colonies were assigned a place "elsewhere" in the "first in Europe and then elsewhere" structure of time' (Bhambra 2009; Chakrabarty 2007: 7; Delanty 2009; Inayatullah and Blaney 2004: 96–103; Outhwaite 2009).

18 In Enlightenment thinking, the mind/body distinction and the elevation of reason separate men from beasts. It is therefore hardly surprising that a colonial grammar casts human nature as uncivilized, that is to say pre-Enlightenment and pre-rational. This implies that those without proper reason, the barbarians who are mired in human nature, are neither sovereign nor masters of themselves and are therefore in need of liberal laws (to intervene and implant the rational) and thus to help the Congolese save themselves from themselves. Pin-Fat explains as follows: 'faculty of reason provides humanity with the possibility of freeing itself from the constraints of certain dispositions such as passions, emotions, basic needs and more importantly, traces of "animalistic urges" such as killing. [...] Reason, in this picture, is what allows us to be "masters" of ourselves; to be sovereign and autonomous and to take ownership of our own actions. In this sense, the human is distinguishable from the animal and indeed, able to master nature' (Pin-Fat forthcoming). See also Dunn (2003).

19 Rapist-soldiers in Bosnia-Herzogovina were also represented as barbaric in relation to the supposedly civilized western Europe (Stanley 1999: 85).

20 See, for instance, Butler (1990); Hall (1996a); Lacan (1977); Mendieta (2003) on answering the call of subject positions. See also Edkins (1999: 84–103) on the Lacanian subject.

21 Many have argued that this is the case in the changing role of the military from warring to peacekeeping and comprehensive approaches to complex emergencies (see Eriksson Baaz and Stern 2011 for an overview of these debates).

22 See Butler (2004b), Hutchings (2008a), Carver (2008a), Pin-Fat (2000) for a critique of the notion of 'human' which allows for some humans to be worthy and others not. Hutchings argues, for example, that 'the only way in which the script of military humanitarian intervention can be sustained is by reproducing differences between people, often most reliably through the naturalizing effects of familiar gendered narratives' (Hutchings 2008a: 3).

23 For instance, the report of the UN Special Rapporteur reads as follows: 'Sexual violence in armed conflicts in the Democratic Republic of the Congo is fuelled by gender-based discrimination in the society at large. [...] The normaliza-

tion and banalization of war-related rape is adding to the inequality and oppression women endure in public and private. The rape crisis associated with war, therefore, cannot be addressed in isolation from gender-based discrimination and violence women encounter in "peace". The war has further reduced women to mere objects that can be raped, tortured and mutilated. Without fundamentally altering gender relations and supporting women's empowerment, high levels of rape will persist, even if stability, the rule of law and democratic, civilian control over the armed forces are established' (Ertürk 2008: 21–2).

24 See also Edkins et al. (1999) on the sovereignty of the self and Butler (2004b) and Pin-Fat (forthcoming) on the question of the human.

25 This is a general report mapping the extent of sexual violence, rather than providing details of specific cases.

26 The stigma attached to male rape is often particularly strong owing to the strong disjuncture between masculinity and victimhood. Being a victim – especially of sexual violence – symbolizes 'failed masculinity', which occupies a position of weakness associated with femininity. Moreover, the stigma is further exacerbated since male rape tends to entail imputing a homosexual identity to the victim (see Lewis 2009).

27 It is difficult to assess the frequency with which men are raped because of the extreme stigma attached to sexual abuse of males and the ensuing reluctance to report such rapes. Nonetheless, it is clear that men are raped in the DRC, as in other conflicts, but it is only recently that such violence has received attention (Sivakumaran 2007). The highest percentage of male victims of sexual violence medical clinics report treating is 6 per cent, while legal clinics report an incidence of 10 per cent (Gettleman 2009a), but the real levels are probably much higher.

28 For testimonies on the latter, see Lewis (2009); Sawyer and Van Woudenberg (2009).

29 At the time of writing, this invisibility is starting to be addressed. Some reports do mention that men and boys are affected by sexual violence. In 2008, the UN Special Rapporteur on violence against women states, for example: 'Women are brutally gang raped, often in front of their families and communities. In numerous cases, men are forced at gunpoint to rape their own daughters, mothers or sisters' (Ertürk 2008: 7; see also UN OCHA 2008). However, as is often the case, the consequences for the male victims forced to rape are not further commented on, and only the raped women are mentioned in discussions of reparations, compensation and justice. Sexual violence is still presented simply as a 'war against women' (Johnson 2009).

30 In contrast to demobilized former child combatants, whose needs are recognized at least discursively (demobilization signifying a discursive move into the victim/survivor camp), those integrated into the army are by default placed in the 'perpetrator' category and thereby lose their rights to reparation, rehabilitation and compensation. Unlike demobilized soldiers, those reintegrated into the army receive no rehabilitation at all. This neglect surely contributes to the violence committed by the army. The increasing levels of sexual violence committed by civilians in recent years are often attributed to an increase in the demobilized combatants (cf. Ertürk 2008; Kippenberg 2009) (many recruited as minors) reintegrated into communities without adequate rehabilitation. Similarly, there is a lack of programmes and initiatives in SSR processes dealing with the special needs and circumstances of previous (especially juvenile) combatants who have been exposed to and forced to commit extreme forms of violence (sexual and other) on civilians. Part of the violence committed by army members must surely be attributed to this lack. At the

moment, the only way in which this issue is addressed is (at best) through 'new orders' that rape is forbidden (see Chapter 4). While these 'new orders' and efforts are commendable and important, they are hardly sufficient to break patterns of learned violent behaviour, especially in the present context of conflicts and unclear command structures.

31 For an exception see Johnson et al. (2010). According to this survey, based on 1,000 villagers in North and South Kivu and Ituri in March 2009, nearly 40 per cent of women and more than 23 per cent of men surveyed reported having suffered sexual assault, mostly rape. Moreover, 41 per cent of female and 10 per cent of male survivors of conflict-related sexual violence said the perpetrator was a woman. However, there are many reasons for treating these high figures with caution, including the length of the questionnaire and the short time allocated for responses, the methodological difficulties involved in getting accurate responses on such sensitive issues, and the conclusions, which contradict other reliable studies, regarding, for instance, the level of civilian abuse and the absence of reporting of the FARDC as perpetrators. Importantly, since the definition of combatant also includes civilians (men or women) abducted and forced to act as sex slaves, it is impossible to decipher much from the 'female perpetrator' category and their actual role in the act.

32 Numerous studies have, for instance, been conducted on women combatants in Sierra Leone and Liberia. For an overview of this literature, see Coulter et al. (2008).

33 For example, the well-deserved focus on the vast devastation caused by widespread conflict-related sexual violence is manifested in (among other things) 'gender and SSR' becoming synonymous with combating sexual and gender-based violence – in particular, rape against civilians. Other gendered aspects, such as the situation of female soldiers as well as the role of women soldiers in violence, are unfortunately neglected in security sector reform efforts (interview material with external actors engaged in SSR in the DRC; see also Eriksson Baaz and Stern 2010). Consequently, gendered interventions have mainly entailed efforts to educate and enlighten male soldiers in order to reconfigure their violent masculinities into responsibilized militarized masculinities, based on an ideal of male soldiers as the disciplined protectors of the civil population with a duty to protect women and children (see, for example, DeGroot 2000).

34 This point has been well addressed elsewhere. See, for example, Moser and Clark (2001).

35 In contrast, consider the 'proper' emotional response offered in the *Guardian*: 'She sought out rapists, and found herself – a slim woman in her late 50s – pointing her camera at men who looked entirely ordinary save their guns, who recounted having carried out five, seven, 20 rapes. "It's hard to keep a record," says one. "We stayed too long in the bush, and that induced us to rape. For an approximate number, maybe 25." While many of the women she interviews are racked with shame, the men who attacked them have none. It is a moral inversion, but typical of attitudes to rape worldwide. Jackson eventually decides that it was seeing these men melt back into the hills that affected her most. "Interviewing the rapists was ghastly," she says, "but the worst moment was when they left. They had just confessed to war crimes, to heinous acts, and I had videotaped it, and then they just sauntered off into the woods. I couldn't help thinking: where are they going, who are their next victims?"' (Cochrane 2008). See also Ahmed (2004); Dauphinée (2007); and Sylvester (2011).

36 Importantly, we do not mean to inflate *our* importance or impact in any of these forums. These questions apply to 'us' as engaged academics, like many

others, and therefore are meant to be posed in a general sense. We revisit this in Chapter 4.

37 For examples of our fledgling attempts at policy recommendations along these lines, see Eriksson Baaz and Stern (2010).

38 We find good company among scholars who, throughout history, have grappled with the problem of understanding and being complicit with that which is considered 'evil'. See, for example, Girard (1995); Hypatia (2003); Neiman (2002); Schott (2003, 2007, 2008); Weil (1965); Wink (1992) for further interrogation of complicity with 'evil'. For other works that address violence from the perspective of the perpetrators, see, for example, Brett and Specht (2004); Hatzfeld (2005); Horwood et al. (2007); Keen (2005); Staub (1989, 2011); Utas (2005); Zimbardo (2008).

39 See Kath Weston for a discussion on the possibility of suspending the moment of representation and categorization in zero time (Weston 2002).

40 Elizabeth Dauphinée explained this in terms of her research among people implicated in war crimes in Bosnia as follows: 'The awareness that there is no possibility of non-violence – that the neighbour of the neighbour is not another neighbour, but also the Other of the neighbour – means that the move towards justice is not a move towards innocence. If even the ethical relation to the Other is haunted by an ineradicable violence, then my capacity to make a non violent decision is radically undermined. This does not mean that I cannot adjudicate – indeed, I do, and I must. What it means, however, is that the place from which I adjudicate is radically different from the system of referents that posits the synchronicity of binary relationships. This means that I cannot pass judgement on Stojan Sokolovic on the cornerstone of my own presumed innocence; instead any judgement of Stojan Sokolovic that I undertake must recognize that he and

I are guilty together – simultaneously guilty – though not interchangeably and not identically. My judgement cannot rest on his excision. Stojan Sokolovic cannot be excised because there is no originary innocence – there is no sphere of non violence – from which to excise him. And thus, there is nowhere for him to go that does not implicate us equally for his crimes' (Dauphinée 2007: 13).

41 When we speak of degendering humanity, we are both drawing from Kath Weston (2002) (who speaks of zero time as the fleeting moment in between significations) and speaking in terms of 'becoming undone' in Butler (2004b) and in the critique of gender in Wiegman (2004). We thus attempt to resist the closing down of the definitions of being human to those who speak in the 'name of humanity'. Being human in this sense is not based on 'common essence' nestled in individual sovereignty, but rather involves 'subjects as produced always already in and through relations with other subjects' (Edkins 2003: 256). Similarly, 'solidarity, in this sense, does not rest on some sort of "essential and universal matter prior to the involvement in relations of power"' (Campbell 1998: 511).

2 'Rape as a weapon of war'?

1 For UN documents referring to Rape as a Weapon of War see, for example, Johnson (2009); MONUC Human Rights Division (2007: 19); UN Action (2007: 3, 5, 2011: 12, 18); UN OCHA (2008: 3, 8, 12).

2 According to Enloe (2000), militarized rape gained visibility during the Yugoslav wars of 1992–95 and the Rwanda conflict of 1994 when the systematic use of rape was raised as a political issue on the international agenda. In 1995, eight Bosnian Serbs were charged with rape in the ICTY, making it the first time rape in war was treated as a separate crime of war. In 1998, the first conviction of rape as a genocidal crime came in the Arusha tribunal in the International Criminal

Tribunal for Rwanda (ICTR) (Enloe 2000: 109, 134, 135, 137, 140). See also Buss (2009) and Henry (2011).

3 See also Buss (2009); Card (1996); Farwell (2004); Kirby (2012); Skjelsbaek (2001, 2010) for overviews/analysis of this framework.

4 See also Kirby (2012) for a discussion of modes of critical explanation of how rape is rendered a weapon of war.

5 We have not addressed the terrain of law specifically in our analysis. See, for instance, Buss (2009); Henry (2011); Park (2007), for such analysis. A good summary of customary international humanitarian law can be found in 'Rule 93: Rape and other forms of sexual violence' by the ICRC. The ICRC also developed 'Practice relating to Rule 93: rape and other forms of sexual violence', which looks at how rape/sexual violence are handled in different types of law (i.e. broken down into treaties, military manuals, UN, international and mixed judicial and quasi-judicial bodies, etc.) (ICRC n.d.-a, n.d.-b). A more in-depth and up-to-date study of the definition of rape in international law can be found in the study conducted by Maria Eriksson (2010).

6 Nodal points (partly) fix the discourse through temporarily stabilizing the meaning of a chain of signifiers in order to impart sense (Laclau and Mouffe 1985: 112; Laclau 1990: 90–2, 109–10).

7 For a slightly different analysis based on the use of 'modes of critical explanation', see Kirby (2012). Kirby identifies 'instrumentality, unreason or mythology as modes of wartime sexual violence' (ibid.: 10).

8 Wodak makes a useful distinction in this context. Drawing on Lemke (1995), she explains: 'discourse is defined on a different more abstract level as text. Discourse implies patterns and commonalities of knowledge and structures, whereas a text is a specific and unique realization of a discourse. Texts belong to genres' (Wodak 2008: 6). Similarly, as Neumann explains, genre carries its own

memory, in the sense that every text relies on its predecessors and carries with it their echoes (Neumann 2008: 69). See also Hansen (2006).

9 Molly MacGregor and Hanna Leonardsson conducted a comprehensive review of UN resolutions, policy documents, UN and NGO reports dealing with conflict-related sexual violence (April–December 2011). Additionally, Hanna Leonardsson conducted a systematic study of articles in the *New York Times* and the *Guardian* ('Reports of rape and sexual violence in war. The Guardian and the New York Times 2003–2011') between September and December 2011 (Leonardsson 2012). This study included those articles that reported on rape in war in general and in the DRC more specifically. She also conducted a more random search of the media via the worldwide web. Together these studies provide the basis for this analysis.

10 Our 'point of saturation' became evident to us as we recurrently recognized the nodal points and their attending representations in different renditions of the main plot.

11 Stuart Hall (drawing on Cousins and Hussain) offers a helpful explanation of Foucault's notion of 'discursive formation' as follows: 'Whenever the discursive event "refers to the same object, shares the same style and supports a strategy [...] a common institutional, administrative or political drift and pattern" [...] then they are said by Foucault to belong to the same discursive formation' (Hall 1997c: 44).

12 See Stern and Zalewski (2009) for a further discussion of generalized storylines.

13 Indeed, as Fairclough explains, 'when searching for a nodal point one would look for signs of over-wording i.e. a proliferation of words that have a presumably similar meaning that is extended across a wide range of policy fields' (Fairclough 2000: 163, cited in Herschinger 2010: 79). The important part

of identifying nodal points *lies in paying attention to what function they play in the discursive formation under study.*

14 For a glimpse of the complexities involved in understanding what can be meant by 'strategy' in other discourses/ fields, consult the well-cited Mintzberg (1992). See also Sloane (2012).

15 'When I refer to the purposes of martial rape, I have in mind its strategic purposes, those appreciable at the level of authority and command. Individual rapists, those who carry out the strategy, may not intend those purposes or be moved by them, just as they may be ignorant of larger purposes served by various orders they implement. There is room [...] at the level of particular acts of rape for many motives' (Card 1996: 9).

16 For accounts of sexual violence in Rwanda, see, for example, Human Rights Watch/Africa and Human Rights Watch Women's Rights Project (1996); Sharlach (1999); and, in the case of Bosnia- Herzegovina, see EC Investigative Mission (1993); Hansen (2001); Niarchos (1995); Skjelsbaek (2010); Stiglmayer (1994); United Nations (1994); Zarkov (1997).

17 For more than a decade, eastern Congo has become infamous as the 'rape capital of the world' and the 'worst place on earth to be a woman' (Viner 2011).

18 For instance, Wallström stated in response to the reports of mass rape in Luvungi, in the North Kivu province of the DRC: 'This terrible incident confirms my general findings during my recent visit to the DRC of the widespread and systematic nature of rape and other human rights violations' (Muscara 2010).

19 Academic usage dates back to at least 1993; see Swiss and Giller (1993).

20 'Whether rape has been explicitly singled out by political and military leaders as a weapon against their op- ponents remains open to question. What is clear is that so far effective measures have rarely, if ever, been taken against such abuses, and that in practice local political and military officers must

have had knowledge of, and generally condoned, the rape and sexual abuse of women, together with the other gross abuses which have so frequently accom- panied the armed conflict in Bosnia- Herzegovina, including the deliberate and arbitrary killing of civilians and the torture and ill-treatment of detainees' (Amnesty International 1993: 4).

21 This study (conducted by Hanna Leonardsson) showed an increased reporting on rape and sexual violence in times of conflict. It also showed that wartime rape or sexual violence has increasingly, in particular from 2006 onwards, been referred to as a weapon, strategy, tactic or tool of war. The term 'weapon of war' was the term most commonly used in the articles but in a few cases rape is also called a 'tactic', 'strategy' or 'tool of war'. For example, rape was called a 'weapon' only three times in 2003 (in both the *New York Times* and the *Guardian*) but in 2011 the study found seventeen references to rape as a 'weapon of war'. In 2010 rape was termed a 'strategy', 'tactic' or 'tool' of war in six cases compared to one single mention in 2003. However, in 2011 the numbers dropped and only one reference to rape as a 'tool' was found.

22 In policy documents the term 'weapon of war' is commonly used to describe both sexual violence and rape. In particular the term is used by NGOs but also by the United Nations in reports and statements (see, for example, Harvard Humanitarian Initiative and Oxfam (2010); Human Rights Watch (2002); Human Rights Watch/Africa and Human Rights Watch Women's Rights Project (1996); Isikozlu and Millard (2010); MONUC Human Rights Division (2007); UN Action (2011); UN OCHA (2008).

23 The 'weapon' terminology figures in UN reports, presidential addresses and General Assembly resolutions in, for example: *Statement by the President of the Security Council. S/PRST/2004/46,* 14 Dec- ember 2004; *Statement by the President*

of the Security Council. S/PRST/2005/25, 21 June 2005; UNFPA and Crossette (2010); and in UN General Assembly Resolutions referring to the Balkans: UN General Assembly (1995, 1996, 1997) and the DRC: UN General Assembly (2006). However, in the well-known UN Security Council Resolutions on Women and Peace and Security, 1820 and 1888, as well as 1882, on Children in Armed Conflict, there are references to rape and sexual violence as a 'tactic' (UNSC 2008, 2009a, 2009b).

24 Human Rights Watch states: '*Rape as a weapon of war* serves a strategic function and acts as an integral tool for achieving military objectives' (Human Rights Watch 2003: 53, emphasis added), and UN Action states: 'UN Action has organized seminars for the academic community and published advocacy articles and OpEds to guide understanding of when sexual violence constitutes a threat to international peace and security, to publicize the use of rape as a tool of political repression, and to explain why it has been war's *"ultimate secret weapon"*' (UN Action 2011: 12, emphasis added). The IRIN/OCHA report states: 'Despite its pervasiveness, rape is often a hidden element of war. Because the use is largely gender-specific and committed by men against women, it is usually narrowly portrayed as being sexual or personal in nature, as a private crime or as a sexual act. Rape, however, is sometimes part of a premeditated political or military strategy. Ignoring the fact that sexual violence against women and girls is used as a combat tactic trivializes what in reality is a war crime' (Horwood et al. 2007: 37–8).

25 See Skjelsbaek (2010) for an overview. See also Leatherman (2011), Buss (2009), Kirby (2012).

26 Three such examples read as follows: 'Gang rape, rape with guns, with torches, with lumps of wood – here in the East of the Democratic Republic of Congo, brutal and systematic rape has become a weapon of war' (Hodgson 2003). 'The United Nations' top relief

official [Jan Egeland] said Tuesday that organized, premeditated sexual attack had become a preferred weapon of war in conflicted parts of Africa, with rapists going unpunished and victims of rape shunned by their communities' (Hoge 2005). 'The UN has called the country [DRC] the centre of *rape as a weapon of war*' (Adetunji 2011).

27 There is confusion between the practitioner level and the resolutions themselves. For example UN Human Rights has a page entitled 'Rape: weapon of war' about Resolution 1820 which actually states that rape is a 'tactic of war.' It is used, for example, in the following: *Statement by the President of the Security Council. S/PRST/2004/46*, 14 December 2004; *Statement by the President of the Security Council. S/PRST/2005/25*, 21 June 2005. In the UN General Assembly Resolutions, in reference to the Balkans: Res. 49/196 (1995), 16; Res. 50/192 (1996), 1–3; Res. 51/115 (1997). In reference to the DRC: Res. 60/170 (2006). In the UN Commission on Human Rights, in reference to the Balkans: C.H.R. Res. 1996/71 (1996); in reference to the DRC: Res. 2005/85 (2005). When referring to the Balkans rape as a weapon of war is often used in the justification of rape as a war crime. Rape as weapon of war is still used in UN rhetoric; for example, it is used to introduce the UNFPA 'State of world population 2010': 'Gender-based violence, including rape, is a repugnant and increasingly familiar weapon of war' (UNFPA and Crossette 2010: Foreword).

28 Compared to the 'rape as a weapon' terminology, which appeared continuously (and in rising numbers) in articles in the *Guardian* and the *New York Times* between 2003 and 2011, the term 'strategy' appeared only in single accounts in 2008, 2009 and 2010. See note 21 for more comparisons of the labelling of rape as a weapon or strategy, tool, tactic.

29 '"Rape in war has been going on since time immemorial," said Stephen

Lewis, a former Canadian ambassador who was the U.N.'s envoy for AIDS in Africa. "But it has taken a new twist as commanders have used it as a strategy of war"' (Kristof 2008).

30 Scholars whose work problematizes the Rape as a Weapon of War discourse, however, have made a point of distinguishing between sexual violence as a strategy and as a tactic, as was explicitly developed in, for instance, Leiby (2009). Wood (2010) explains that strategy and tactic can be seen as employing two logics: strategy implies 'sexual violence promoted by group leadership against specific populations' (sexual torture of political prisoners, cleansing, collective punishment, low-cost reward to troops) because they perceive it as effective. A top-down logic reigns. Tactic: 'combatants/small units practice sexual violence that is not ordered because they perceive it as effective [... it] comes from low levels of [the] command structure so it's bottom-across logic; for it to spread the group's military hierarchy must be tolerated explicitly or implicitly' (Wood 2010: 316–18). We will return to these points in the next chapter.

31 Examples of weapon regulation for other forms of weapons can be seen, for instance, in the Convention on Certain Conventional Weapons (CCW), which states: 'Basing themselves on the principle of international law that the right of the parties to an armed conflict to choose methods or means of warfare is not unlimited, and on the principle that prohibits the employment in armed conflicts of weapons, projectiles and material and methods of warfare of a nature to cause superfluous injury or unnecessary suffering' (United Nations 1980). In addition, international law concerning weapons of mass destruction from 1969 states that 'Existing international law prohibits, irrespective of the type of weapon used, any action whatsoever designed to terrorize the civilian population' (Institute of International Law 1969: §6).

32 Interviews with UN personnel, 2009/10.

33 Ibid. It seems, nonetheless, that this shift in rhetoric was not adhered to. In 2011 the UN Action Initiative continued to use the term 'rape as a weapon of war' in its own progress report (UN Action 2011: 12, 18).

34 Resolution 1820 states: '[...] sexual violence, when used or commissioned as a tactic of war in order to deliberately target civilians or as a part of a widespread or systematic attack against civilian populations, can significantly exacerbate situations of armed conflict and may impede the restoration of international peace and security ...' (UNSC 2008: 3). Resolution 1888: '*Reaffirms* that sexual violence, when used or commissioned as a tactic of war in order to deliberately target civilians or as a part of a widespread or systematic attack against civilian populations, can significantly exacerbate situations of armed conflict and may impede the restoration of international peace and security; ...' (UNSC 2009b: 3).

35 For instance: 'The raping of women and girls is an all-too-common *tactic* in the war raging in Congo's eastern jungles involving numerous militia groups' (Lacey 2004).

36 For further discussion of the differentiation between sexual violence as a strategy and as a tactic see Leiby (2009) and Wood (2010).

37 For example, Johnson (2009) describes rape and sexual violence as a weapon, a tactic and a strategy, while Wallström, in her statement to the Security Council (2010b), refers to it only as a tactic. Also, in reports such as *UN Action against Sexual Violence in Conflict* (UN Action 2007), the three terms can be found to describe sexual violence and rape.

38 See Schott's critique of Card for a further discussion of evil as it is connected to intent (Schott 2004, 2011).

39 The politics of establishing guilt are certainly complicated. For instance: '"There were several months of crimes,

killings, lootings but there was a peak of rapes in a few days," he [Luis Moreno-Ocampo, prosecutor at the ICC] said. "This will be our biggest challenge, to prove that someone was authorizing them or giving instructions." The case was complex, he said, because his office would "not prosecute the rapists themselves," but the person or people issuing the orders or organizing the campaign' (Polgreen and Simons 2007).

40 As we shall see in Chapter 3, military institutions seldom (if ever) work according to the ideals of discipline, hierarchy and control.

41 For an analysis of the politics of different representations of how rape should be understood in the former Yugoslavia, see Hansen (2001). For instance, she states: 'The large-scale raping of Bosnian women – commonly suggested to be as many as 20,000 – and the perceived inability of the Bosnian men to provide protection were part of Serbian attempts to constitute the entire Bosnian nation as humiliated, inferior, weak and feminine. However, the precise construction of this nationalized-gendered subject, and its implications, were by no means uncontested. Different groups offered competing understandings of the meaning and causes of the rapes as well as which policies should be undertaken towards them. The goal of this article is to explore the dynamics involved in these constructions of the Bosnian mass rapes as a possible security problem' (ibid.: 56).

42 Rape has been treated as a crime in war as far back as 1919, although, as Henry explains, 'although wartime rape has been repeatedly condemned as the "worst of crimes" throughout history in political rhetoric, in practice these crimes have been very much neglected, disregarded, denied and downplayed' (Henry 2011: 6).

43 This distinction complicates the question of the evilness of particular acts of rape. See Schott (2011: 10 and 2004).

44 For instance, note the wording in the following statement from Human Rights Watch: 'The widespread incidence of rape accompanied this increase in overall violence against groups previously immune from attack. "Rape was a strategy", said Bernadette Muhimakazi, a Rwandan women's rights activist. "They *chose to rape.* There were no mistakes. During this genocide, everything was organized. Traditionally it is not the custom to kill women and children, but this was done everywhere too"' (Human Rights Watch/Africa and Human Rights Watch Women's Rights Project 1996: 23, emphasis added).

45 See Zimbardo (2008) and Staub (2011) for a discussion of how allegiances to a group/nation/social order enable the suspension of individual codes of ethics.

46 Véronique Pin-Fat explains as follows: 'faculty of reason provides humanity with the possibility of freeing itself from the constraints of certain dispositions such as passions, emotions, basic needs and more importantly, traces of "animalistic urges" such as killing. We might say then that reason is what allows us to overcome animalistic aspects of our nature and thereby, of most pertinence here, provide the possibility of ethics. We can control ourselves and refrain from simply acting on "passions" such as hate, fear, revenge, lust and so on. Reason, in this picture, is what allows us to be "masters" of ourselves; to be sovereign and autonomous and to take ownership of our own actions. In this sense, the human is distinguishable from the animal and indeed, able to master nature' (Pin-Fat forthcoming).

47 The responsibility to protect (women) against sexual violence in conflict is part of international law and found in, for example, First Geneva Convention, 1949: Art. 3, 27; Fourth Geneva Convention, 1949: Art 3, 27; Second Geneva Convention, 1949: Art. 3; Third Geneva Convention, 1949: Art. 3. The notion that sexual violence can constitute

war crimes, crimes against humanity or acts of genocide is found in, for example, Protocol (II), 1977; Rome Statute, 1998: Art 7; UNSC (2000: Art. 11, 2008). The ending of impunity and the exclusion of crimes of sexual violence from deals of amnesty is emphasized in many resolutions, for example UNSC (2000: Art. 11, 2008: Art. 4, 2009b: Art. 4).

48 In addition the notion of command responsibility, or the obligation to prosecute the persons committing or ordering to be committed any breaches, is present in, among others, First Geneva Convention, 1949: Art. 49; Fourth Geneva Convention, 1949: Art. 146; Rome Statute, 1998: Art. 28; Second Geneva Convention, 1949: Art. 50; Third Geneva Convention, 1949: Art. 129, and specifically pointing to the responsibility of commanders to prevent breaches of the Geneva Conventions in Protocol (I), 1977: Art. 87.

49 Dennis Mukwege is the chief gynaecologist and medical director at the Panzi hospital in Bukavu, DRC.

50 The notion of unavoidability is (not surprisingly) particularly dominant in policy discourse, although it also appears in academic discourse.

51 For a good analysis of the politics of responsibilization, see Hansson (forthcoming).

52 See also research and policy reports that come out of the Woman Stats Project (n.d.), where links between gender inequality in society and propensity for warring more generally are rendered explicit.

53 For instance: 'Throughout the world, sexual violence is routinely directed against females during situations of armed conflict. This violence may take gender-specific forms, like sexual mutilation, forced pregnancy, rape or sexual slavery. Being female is a risk factor; women and girls are often targeted for sexual abuse on the basis of their gender, irrespective of their age, ethnicity or political affiliation' (Human Rights Watch/Africa and Human Rights Watch Women's Rights Project 1996). 'It is now more dangerous to be a woman than a soldier in modern wars' (Major General Patrick Cammaert, former Deputy Force Commander, MONUC, quoted in UNDP 2008). 'Other Rwandans characterized the choice of violence against women in the following ways: "It was the humiliation of women"; or "It was the disfigurement of women, to make them undesirable, used"; or, finally, "Women's worth was not respected"' (Human Rights Watch/Africa and Human Rights Watch Women's Rights Project 1996).

54 For a discussion of rape as a war on women in newspaper articles see: Elahi (2007); Herbert (2006, 2009); Hodgson (2003); Kahorha (2011); McGreal (2008). For reports referring to wartime rape as a war on women see: Amnesty International (2008); Human Rights Watch (2002: 1, 2009b: 15); Ohambe et al. (2005: 48).

55 Many feminist theorists have dealt with this difficult question connecting wartime rape to patriarchy in different ways (Barstow 2000; Brownmiller 1975; Card 1996; Seifert 1996; Stiglmayer 1994). Broadly speaking, a dominant line of thinking posits that sexual violence in conflict is largely a result of men's domination over women and/or motivated by the desires of man to exert dominance over women in patriarchal societies. Wartime rape is thus similar to rape in peacetime. As Brownmiller explains, wartime rape is understood as a result of male combatants who 'vent their contempt for women' (Brownmiller 1975: 32). According to Seifert (1996: 37), a central aspect of wartime rape is 'hatred of women in general'. The brutality of some rape cases has an important function in these explanations, particularly for Seifert. In citing the 'quasi-ritualistic' nature of the mutilation of female body parts of some of the crimes in the Balkans, she contends: 'only a hatred of femininity as such can account for that specific kind of violence' (Sabalic and Seifert 1993, cited in Seifert 1996: 37). Ertürk explains: 'The

normalization and banalization of war-related rape is adding to the inequality and oppression women endure in public and private. The rape crisis associated with war, therefore, cannot be addressed in isolation from gender-based discrimination and violence women encounter in "peace". The war has further reduced women to mere objects that can be raped, tortured and mutilated. Without fundamentally altering gender relations and supporting women's empowerment, high levels of rape will persist, even if stability, the rule of law and democratic, civilian control over the armed forces are established' (Ertürk 2008: 22).

3 The messiness of warring

1 We recognize the differences already mentioned in Chapter 2 between strategy (alluding to more general plans of how to employ military power to achieve objectives formulated by the leadership) and tactic (alluding to the means used to implement the strategy, or 'innovations' at the lower levels of the command structure) (see also Wood 2010 and Leiby 2009).

2 The Congolese army consists of the former government forces and several armed groups, which have signed peace agreements. The first round of military integration after the peace accord in 2002 included the signatories of the Global and All-inclusive Agreement (*inter alia* the government forces FAC, MLC, RCD and the Mai-Mai). However, military integration turned out to be a continuous process, including armed groups that were created and mobilized after the 2002 peace accord. With the signing of new peace accords with remobilized armed groups, there has been a constant addition of new armed units to be integrated over the years. Hence, the army has been undergoing constant reorganization, involving the breaking up and forming of new brigades. This integration has often involved units and fighters that have already been integrated in an earlier

phase and then abandoned the process, only to join again (Eriksson Baaz 2011; Verweijen 2013a).

3 While our own research did not feature any direct accounts of ordered rape, it demonstrated a permissive climate in relation to sexual violence. Rationales behind such permissiveness often evoked the need for 'substitution' for soldiers whose sexual urges were not satisfied in the 'normal' way (Eriksson Baaz and Stern 2009). In line with the underlying assumptions of the 'Sexed' Story discussed in Chapter 1, commanders, especially those interviewed at the beginning of the research, often contended that rape is 'very difficult to stop'. While neither officers nor soldiers ever explicitly spoke of rape as strategic, such attitudes could be read as an implicit authorization, which could be portrayed as being strategic. As discussed in the previous chapter, most proponents of the 'weapon of war' narrative embrace a broad understanding of strategy (cf. Allen 1996; Card 1996; Skjelsbaek 2001) where the strategicness or 'weapon'-ness of rape does not presuppose orders from the chain of command. However, as we will argue later in the chapter, the strategicness of rape can also be located not mainly in its functionality as a tactic to intimidate and punish local populations, but in the role it plays in safeguarding commanders' own positions.

4 While 'winning the hearts and minds' is a concept with a long and varied history (see Dickinson 2009), the concept has become increasingly popular, and occupies a central role in US interventions. It is also inscribed in the US Army and Marine Corps Counter-insurgency manual: 'Protracted popular war is best countered by winning the "hearts and minds" of the populace' (ibid.).

5 The massacre was part of a counter-insurgency against the Muslim Brotherhood. According to Van Creveld, this counter-insurgency attack, which seri-

ously weakened the Brotherhood and left the population in a state of shock and horror, was successful given its political aims; it saved the regime, scattered the insurgent and forced the population into obedience.

6 We draw great inspiration from having delved into relevant literature in the fields of the sociology of the military specifically and the sociology/anthropology of violence more generally.

7 Here, one's position is maintained through the distribution of these resources in complex clientist networks, both inside and outside the military institution. In addition, the politics of integration has contributed to constant defections whereby armed units break away and mobilize again, often with the hope of striking a better deal in military integration next time around (Verweijen 2013a; Eriksson Baaz 2011; Vlassenroot and Raeymaekers 2009).

8 It is also the implicit narrative in the more popular activist 'conflict/rape minerals' discourse, in which rape is portrayed simply as an outcome of armed groups' involvement in the illegal mineral trade. (A familiar argument is that 'our' cell phones produce rape in the Congo; see Autesserre 2012 for an analysis of this discourse).

9 For a similar conclusion in relation to the FAPC, see Titeca (2011).

10 See also Bourke (1999: 57–91) on military training.

11 See Foucault (1977) on the 'docile bodies' produced in the military.

12 Even the 'cold' parts of the military organization (i.e. the MoD and Chief of Staff) are sometimes rather an 'organized anarchy' where 'information [...] becomes lost in the system, directed to the wrong people, or both' and where, during times of crises, 'the wrong people may try to solve a problem because of their prowess at bureaucratic gamesmanship, or the right people [...] may be overlooked or sent elsewhere' (Sabrosky et al. 1982: 142, cited in Soeters et al. 2006: 246). Hence,

the efficient, rational military remains an ideal, which has turned out to be difficult to attain in practice in militaries all over the globe, on the battlefield and elsewhere.

13 Brighton's reading here emphasizes how Clausewitz's writings reflect both his own experience as a soldier and the outcome of a moment of violent transformation.

14 It is for this reason that 'classic' officer training emphasizes the need for commanders to act as role models for the troops, through sharing (rather than avoiding) soldiers' hardships, and in so doing demonstrating the military virtues of discipline and sacrifice.

15 While our research did not provide any accounts of directly ordered sexual violence, other research, which addresses other military units and focuses on violence against civilians more generally, has provided some examples of commanders ordering acts of violence against civilians (Verweijen 2013a). However, Verweijen's research points to this being quite rare.

16 Particularly so in companies, brigades and battalions, less so at lower levels (platoons and sections). See Verweijen (2013a).

17 While ex-FAC soldiers have been promoted in a system that, while certainly fallible, was based on competence and merit, many of the newly integrated soldiers have little or no training but still hold a high rank, something which causes resentment and diminishes the legitimacy of the commanders in the eyes of the troops.

18 The parenthetical insertion of imagined corruption is made to highlight another important point, namely endemic and normalized mistrust. While corruption in the justice system in the DRC (as in other areas) is real, the general mistrust created by decades of mismanagement and corruption has created a generalized and widespread mistrust of authorities. This mistrust

will in itself constitute a problem, even if major changes take place towards an independent and incorrupt judiciary. Regaining trust will most surely take a very long time.

19 This quotation also appeared in Eriksson Baaz and Stern (2008).

20 Only a minority of the ex-combatants interviewees in Sierra Leone, for instance, stated that they had ever heard a commander give an order to rape.

21 For an interesting analysis drawing on Collins's concept of 'forward panic', see Higate (2012a).

22 According to Collins (2008: 100): 'In the Nanking massacres, the initiating event was the Japanese commander's order to kill the Chinese prisoners, out of practical considerations of guarding such large numbers; this in turn unleashed a moral holiday, emotionally fuelled by the tension of the Japanese troops in their prior campaign, now confronted with the total collapse of resistance from their enemy. The frenzy of destruction went beyond any rational scorched earth policy or exemplary terror, and is best seen as an unusually prolonged forward panic.'

23 This quotation also appeared in Eriksson Baaz and Stern (2010).

24 See Eriksson Baaz and Stern (2009) for further discussion of this point.

25 Moreover, many survivors of sexual violence in the DRC in their testimonies speak of aggressors who are (sometimes seriously) intoxicated.

26 See Amnesty International (2012).

27 Interviews with UN and Congolese army staff. For official sources, see also Stearns (2011) and (BBC 2011a).

28 See note 16.

29 While the acts were clearly recognized as 'wrong', the retelling of the acts was not accompanied by sentiments of guilt or remorse. This apparent absence of guilt should not be interpreted (as it sometimes is) as a manifestation of a lack of morality or capacity to distinguish 'right' from 'wrong'. Rather, it could

be understood in relation to masking, in a symbolic sense; a common trait of soldiering which implies moving oneself from reality, stepping into a state of non-normativity. See Utas (2008) for a discussion on 'masking'.

4 Post-coloniality, humanitarian engagement

1 See, for example, Stearns (2009a).

2 The Netherlands, one of the first foreign countries to draw attention to the issue by sending their minister of development cooperation to the DRC in 2006, cared so much that they ordered their embassy staff to bring up the issue of sexual violence in all meetings with the government and other donors, irrespective of the content of the meeting.

3 As Hall points out here, the 'post-colonial' is similar to other 'posts': 'It is not only "after" but "going beyond" the colonial, as post-modernism is both "going beyond" and "after" modernism, and post-structuralism both follows chronologically after and achieves its theoretical gains "on the back of" structuralism' (Hall 1992: 253). As Simon Gikandi (1996: 14–15) puts it, post-colonial theory is 'one way of recognizing how decolonized situations are marked by the trace of the imperial pasts they try to disavow'. It can be understood as a 'code for the state of undecidability in which the culture of colonialism continues to resonate in what was supposed to be its negation'.

4 The study 'Reports of rape and sexual violence in war. The Guardian and the New York Times 2003–2011', conducted by Hanna Leonardsson (September–December 2011), systematically collected and studied articles in the *New York Times* and the *Guardian* that reported on rape in war in general and in the DRC more specifically, together with a more random search of the media via the worldwide web.

5 Since 2007, in articles from the *New York Times* and the *Guardian*, the account

of rape as a weapon of war has been accompanied by an increasing reference to rape as a tactic or strategy of war. See, for example, Herbert (2009); Kristof (2008); McGreal (2007); Smith (2010); Wallström (2010a).

6 For an interesting analysis of the emergence of rumours of cannibalism in the Ituri that later enjoyed a lot of attention in Western press, see Pottier (2007). As Pottier argues, these rumours must primarily be understood as a politically driven metaphor of the violence and suffering experienced in the conflict.

7 For stories of cannibalism, see, for example, Astill (2003); Isango (2003); Left (2005); New York Times (2003).

8 Gettleman here cites the prize-winning chief surgeon of the Panzi hospital, Denis Mukwege. Our criticism of this passage in the article has been attacked by some who argued that the critique is unfair since it is Mukwege who makes the comparison with gorillas. However, we believe that this is to miss the point. Our analysis and critique is not intended to analyse authors/utterers and their intentions, but discourse; how we make meaning of the DRC warscape and its violence. Hence, the question of who made the statement in the first place is irrelevant and our intent is not to label certain individuals (in this case Gettleman) as racists. However, this does not mean that Gettleman has no responsibility for his text; he clearly has a choice in what parts to cite in the (presumably long, given our knowledge of Mukwege) interview on which the article is partly based. Moreover, making racist claims by references to 'a national' through the argument 'they say it themselves' is a classic strategy and does not, in any way, alter racist representations (the analogy combatants = gorillas in this case).

9 Moreover, it should be remembered that the use of various 'irrational/ magical' objects and practices in the belief that these can enhance fortune in combat (and attributing misfortune to

the lack thereof) is not akin to modernity and modern warfare. See, for example, Emmens (2010); Milbank (2000); Mirage Men (2010); Royal Air Force Museum (n.d.).

10 Conversation with observer in the DRC, November 2010.

11 As Gustave le Bon (1879), *La Psychologie des Foules* (quoted in McClintock 1995: 54), argued, 'All psychologists who have studied the intelligence of women, as well as poets and novelists, recognize today that they represent the most inferior forms of human evolution and that they are closer to children and savages than to an adult, civilized man.'

12 Eve Ensler, 'playwright, performer, and activist, is the author of The Vagina Monologues, translated into over 48 languages and performed in over 140 countries [...] Ms. Ensler's experience performing The Vagina Monologues inspired her to create V-Day, a global movement to stop violence against women and girls. She has devoted her life to stopping violence, envisioning a planet in which women and girls will be free to thrive, rather than merely survive [...] Today, V-Day is a global activist movement that supports anti-violence organizations throughout the world. [...] In late 2010, V-Day and UNICEF, in partnership with Panzi Foundation, will open [*read: opened*] the City of Joy, a special facility for the survivors of sexual violence in Bukavu, Democratic Republic of Congo (DRC)' (V-Day n.d.).

13 In debates on the vices and virtues of Western modernity, African and other 'non-Western' cultures have functioned as the Other to which Western modernity is compared. The relation to modernity in Europe has always been marked by ambivalence – ambivalence that also has been reflected in representations of the colonized Other. The colonized Other has functioned as an object of desire for critics who have argued that Western civilization fills people with unnatural wants and ambitions and seduces them from

their true nature. The positioning of the colonized on the lowest rung of the evolutionary ladder has not only been used as an illustration of barbarism and chaos in Western texts. The colonized Other has, at the same time, also functioned as an object of desire (Gikandi 1996; Hall 1992, 1996a, 1997a; Pieterse 1992b; Young 1995). The 'savage debate' during the seventeenth and eighteenth centuries is perhaps the best-known example of how battles between different visions of the future of Europe have been played out on the non-western Other, resulting in the division of the savage into the noble and the ignoble, existing side by side, reflecting different visions and programmes. The Noble Other, leading a simple and innocent life untouched by the vices of Europe, served a nostalgic and political purpose, a reminder of the past, but also of the possible futures of Europe.

14 Spivak locates the problem not in the inability of the subaltern to speak, but in the unwillingness of the culturally dominant to listen. Here it is not only the British but also the Indian elites who oppress the subaltern subject. She challenges the simple division between colonizers and colonized by introducing the brown woman as a category oppressed by both. The story of Bhuvaneswari Bhaduri, who committed suicide in 1926 and who did all in her power to contest the representation of her suicide as sati resulting from illicit love and pregnancy by taking her life while she was menstruating, is a powerful illustration of the systematic silencing of the subaltern. Such silencing occurs not only in narratives of imperialism, but also by those who are assumed to occupy the same position of subalterneity: her own female relatives.

15 We have participated in such events partly at workshops and conferences where women survivors of rape have been invited to tell their stories. We have also participated as facilitators when politicians or representatives from donor organizations have visited the DRC and wanted to meet and talk to victims of sexual violence.

16 A good (official source) example of this can be seen in the reporting of Margot Wallström's visit to Walikale in 2010, to meet the rape victims of the massive rape (303 people, of which 16 were men), which also featured pillage (923 homes pillaged), abductions (116 people), in the documentary *Resolution* by Marika Grisel (2011). As in many other instances, the women who speak here talk about their lack of the basic means of survival (land to cultivate, clean water, etc.) but this message is subsumed by the main narrative of rape.

17 Importantly, we do not mean to imply that others more steeped in colonial logics than ourselves deserve our righteous contempt for their selective listening. We have ourselves been guilty of picking through our respondents' stories to 'get to' the rape part, because rape was the *focus* of our research. Furthermore, as we discussed in Chapter 1, the dominant frames of understanding condition what we hear as well as how we render sense of what we hear, even when we strive to listen differently. We discuss this further in Chapter 5.

18 Again, we are not immune from this (self-)critique. In many different contexts, our research has imbued us with an air of the extraordinary because of its subject matter and our encounters with 'rapists' in the supposedly 'worst place in the world to be a woman' (*Daily Mail*, 12 May 2011).

19 These organizations are of diverse nature. While some are simply 'one man/woman organizations', primarily driven by the objective of using the new funding opportunities for their own personal benefit, many have a firm base in the communities in which they work and do a remarkable job.

20 Interviews with UN staff in the DRC, 2010 and 2011.

21 While many of these resources have been channelled into service pro-

vision to rape survivors, the specific focus on sexual violence (in relation to other violence) is also reflected in violence-prevention-related interventions. For example, while the population (particularly of the eastern parts of the country) is informed that women have the right not to be raped (through posters, radio emissions, banderoles and other means of communication which warn potential perpetrators/viewers/readers of the grave legal consequences), similar communications in relation to other forms of violence are quite absent.

22 Interviews with international and national NGOs in Kinshasa and Bukavu, 2010 and 2011. However, given the sensitive nature of criticism of intervention in the field of sexual violence, this critique was mostly expressed in off-the-record interviews (see also Autesserre 2012).

23 For an official source, see, for example, York (2010).

24 A common practice in many areas of the DRC when a man dies is that the family of the deceased comes to the wife/widow and claims property, sometimes even the house itself. While the woman – if the marriage was registered by the local authorities – has the legal right to the inheritance, many are not aware of their legal rights. Moreover, most women lack the financial means to access legal support and take the case to the tribunal. However, even if they are aware and have the necessary funds, many choose not to take the case to justice owing to the pressure of customary beliefs and practices and fears of what taking the case to court might bring in the form of actions from the in-laws that could hurt the well-being of the children.

25 Interviews with health personnel in South Kivu. In some instances, a rape victim (or even attempted rape victim) status allowed for free or preferential treatment for all sorts of injuries and health problems, not just rape-related ones. For instance, according to one account, an elderly woman who had a gunshot wound was encouraged to invent an official story that she was shot while trying to escape a rape (which was not the case) in order to receive free treatment from the earmarked funding for rape victims (interview with a researcher conducting research on health-related issues in the eastern DRC).

26 Interviews with representatives of local organizations providing support to rape survivors in South Kivu.

27 MSF France, which was responsibile for one of the camps, finally decided to withdraw on the grounds that they were unable to fulfil their task of 'protecting human dignity' and because 'the amputees' recovery process' was 'jeopardized by the sheer number of visitors' (MSF, cited in Polman 2011: 62).

28 Law 0018-0019.

29 This covers both violent, forced sexual intercourse and what in the USA is defined as statutory rape: that is, consensual sex between two people where at least one of them is under the legal age.

30 Interviews with local organizations in South Kivu. For a telling and longer account of one such case, see Sudetic (2011).

31 During a meeting with Margot Wallström in Stockholm in June 2010, hosted by Sida, we raised the problem of the single focus on sexual violence, criticizing UN action in this regard, pointing out the perverse consequences manifested in the commercialization of rape and arguing (as we have here) that the interventions problematically isolate sexual violence from, and take away the focus from, other problems identified and prioritized by Congolese women organizations. In her response Margot Wallström defended the UN approach, categorically denied the problem of commercialization and contended that she was convinced of the opposite: that the focus on sexual violence attracts attention also to other issues identified by Congolese women.

32 As Christine Sylvester argued some time ago, there is a tendency within postcolonial studies to not 'concern itself with whether the subaltern is eating' (Sylvester 1999: 703).

33 See the website 'Good intentions are not enough': goodintents.org/.

34 We mean representing in both of the following ways: speaking for, in the sense of political representation, and representing, speaking about, portraying, the woman as she really is (see Spivak 1988).

5 Concluding thoughts

1 While the 'Substitution Theory' (discussed with the 'Sexed' Story in Chapter 1) does not necessarily entirely lack officially and morally acceptable recommendations, such as increased disciplinary measures to curtail (natural male) behaviour, its policy recommendations are rather associated with providing a 'natural', non-violent outlet for the fulfilment of male sexual desires (through increased access to prostitutes, wives or girlfriends in deployment areas).

2 Higate (2012a: 26) argues in relation to his analysis of the Personal Security Detail (PSD) in the much-publicized 2007 shootings in Baghdad's Nisour Square, where seventeen Iraqis were massacred: 'the question of whether or not [the perpetrators] should be punished falls outside of a sociological remit'.

3 As Spivak has suggested, reading fiction is useful here 'as a way of remembering again how to imagine', 'put[ting] ourselves in the protagonist shoes, suspend belief, and let ourselves be surprised by the twists and turns of the plot'.

Bibliography

Abrahamsen, R. (2004) 'The power of partnerships in global governance', *Third World Quarterly Journal of the Mythic Society*, 25(8): 1453–67.

Adetunji, J. (2011) 'Forty-eight women raped every hour in Congo, study finds', *Guardian*, 12 May, www.guardian.co.uk/world/2011/may/12/48-women-raped-hour-congo.

Agamben, G. (1998) *Homo Sacer: Sovereignty, Power and Bare Life*, Stanford, CA: Stanford University Press.

Ahmed, S. (2004) *The Cultural Politics of Emotion*, Edinburgh: Edinburgh University Press.

Ali, S. (2007) 'Feminism and postcolonial: knowledge/politics', *Ethnic and Racial Studies*, 30(2): 191–212.

Alison, M. (2007) 'Wartime sexual violence: women's human rights and questions of masculinity', *Review of International Studies*, 33(1): 75–90.

Allen, B. (1996) *Rape Warfare: The Hidden Genocide in Bosnia-Herzegovina and Croatia*, Minneapolis: University of Minnesota Press.

Amnesty International (1993) *Bosnia-Herzegovina: Rape and sexual abuse by armed forces*, London: Amnesty International Publications.

— (2008) *Democratic Republic of Congo: North Kivu no end to war on women and children*, London: Amnesty International Publications.

— (2012) 'Annual Report 2012. Democratic Republic of the Congo', www.amnesty.org/en/region/democratic-republic-congo/report-2012.

Astill, J. (2003) 'Congo cannibalism claim provides first challenge', *Guardian*, 11 March 2003, www.guardian.co.uk/world/2003/mar/11/congo.jamesastill.

Autesserre, S. (2012) 'Dangerous tales: dominant narratives on the Congo and their unintended consequences', *African Affairs*, published online 9 February, doi: 10.1093/afraf/adr080.

Aylwin-Foster, N. (2005) 'Changing the army for counterinsurgency operations', *Military Review*, November/December, pp. 1–15.

Barkawi, T. and S. Brighton (2011) 'Powers of war: fighting, knowledge, and critique', *International Political Sociology*, 5(2): 126–43.

Barstow, A. L. (2000) 'Introduction', in A. L. Barstow (ed.), *War's Dirty Secret. Rape, prostitution, and other crimes against women*, Cleveland, OH: Pilgrim Press.

Bartone, P. T., B. H. Johnsen, J. Eid, W. Brun and J. C. Laberg (2002) 'Factors influencing small-unit cohesion in Norwegian Navy officer cadets', *Military Psychology*, 14(1).

Basham, V. (2009) 'Effecting discrimination: operational effectiveness and harassment in the British armed forces', *Armed Forces and Society*, 35(4): 728–44, doi: 10.1177/0095327X08324762.

Bass, B. M., B. J. Avolio, D. I. Jung and Y. Berson (2003) 'Predicting unit performance by assessing transformational and transactional leadership', *Journal of Applied Psychology*, 88(2): 207–18.

BBC (2010) 'UN official calls DR Congo "rape capital of the world"', *BBC News*, 28 April, news.bbc.co.uk/2/hi/8650112.stm.

— (2011a) 'DR Congo army commander "led mass rape" in Fizi', *BBC News*, 19 January, www.bbc.co.uk/news/world-africa-12205969.

— (2011b) 'DR Congo colonel Kibibi Mutware jailed for mass rape', *BBC News*, 21 February, www.bbc.co.uk/news/world-africa-12523847.

Benhabib, S., J. Butler, D. Cornell and N. Fraser (1995) *Feminist Contentions: A Philosophical Exchange*, New York: Routledge.

Bhabha, H. K. (1994) *The Location of Culture*, New York: Routledge.

Bhambra, G. K. (2009) 'Postcolonial Europe, or understanding Europe in times of the postcolonial', in C. Rumford (ed.), *The Sage Handbook of European Studies*, London: Sage, pp. 69–86.

Boshoff, H. (2007) 'Demobilisation, Disarmament and Reintegration in the Democratic Republic of Congo', *African Security Review*, 16(2): 59–62.

Bourke, J. (1999) *An Intimate History of Killing: Face-to-Face Killing in Twentieth-century Warfare*, London: Basic Books.

— (2007) *Rape: A History from 1860 to the Present*, London: Virago.

Braidotti, R. (2006) *Transpositions: On nomadic ethics*, Cambridge: Polity.

Brassett, J. and D. Bulley (2007) 'Ethics in world politics: cosmopolitanism and beyond?', *International Politics*, 44: 1–18, doi: 10.1057/palgrave.ip.8800155.

Braudy, L. (2003) *From Chivalry to Terrorism: War and the changing nature of masculinity*, New York: Alfred A. Knopf.

Brett, R. and I. Specht (2004) *Young Soldiers: Why they choose to fight*, Boulder, CO: Lynne Rienner.

Brighton, S. (2011) 'Three propositions on the phenomenology of war', *International Political Sociology*, 5(1): 101–5.

Broch-Due, V. (ed.) (2005) *Violence and Belonging: The quest for identity in postcolonial Africa*, Abingdon: Routledge.

Brownmiller, S. (1975) *Against Our Will: Men, women and rape*, New York: Ballantine.

Burchell, G. (1996) '"Liberal government and techniques of the self." Foucault and political reason: liberalism, neoliberalism and rationalities of government', in A. Barry, T. Osborne and N. Rose (eds), *Foucault and Political Reason. Liberalism, neoliberalism and rationalities of government*, Chicago, IL: University of Chicago Press.

Bureau of Democracy, Human Rights and Labor (2009) *2009 Human Rights Report: Democratic Republic of the Congo*, Washington, DC: Department of State.

Burke, A. (2007) *Beyond Security, Ethics and Violence: War against the other*, London and New York: Routledge.

Burton, A. M. (1990) 'The white woman's burden: British feminists and the Indian women, 1865–1915', *Women's Studies International Forum*, 13(4): 295–308.

Buss, D. E. (2009) 'Rethinking "rape as a weapon of war"', *Feminist Legal Studies*, 17(2): 145–63, doi: 10.1007/s10691-009-9118-5.

Butler, J. (1990) *Gender Trouble: Feminism and the subversion of identity*, London: Routledge.

— (2004a) *Precarious Life: The Powers of Mourning and Violence*, London and New York: Verso.

— (2004b) *Undoing Gender*, New York: Routledge.

— (2005) *Giving an Account of Oneself*, New York: Fordham University Press.

— (2009) *Frames of War: When Is Life Grievable?*, London: Verso.

Caforio, G. (2006) 'Military officer education', in G. Caforio (ed.), *Handbook of the Sociology of the Military*, New York: Springer.

Campbell, D. (1998) 'Why fight: humanitarianism, principles, and post-structuralism', *Millennium – Journal of International Studies*, 27, doi: 10.1177/03058298980270031001.

Campbell, D. and M. J. Shapiro (eds) (1999) *Moral Spaces: Rethinking ethics and world politics*, Minneapolis: University of Minnesota Press.

Card, C. (1996) 'Rape as a weapon of war', *Hypathia*, 11(4): 5–18.

Carpenter, C. R. (2006) 'Recognizing gender based violence against civilian men and boys in conflict situations', *Security Dialogue*, 37(1): 83–103.

Carroll, R. (2005) 'Eight years of darkness', *Guardian*, 31 January, www.guardian.co.uk/world/2005/jan/31/gender.uk/print.

Carver, T. (2008a) 'Real construction through metaphorical language: how animals and machines (amongst other metaphors) makest (hu)man what "he" is', in T. Carver and J. Pikalo (eds), *Political Language and Metaphor: Interpreting and Changing the World*, London: Routledge, pp. 151–64.

— (2008b) 'The machine in the man', in J. Parpart and M. Zalewski (eds), *Rethinking the Man Question: Sex, gender, and violence in international relations*, London: Zed Books.

Chakrabarty, D. (2007) *Provincializing Europe: Postcolonial Thought and Historical Difference*, Princeton, NJ: Princeton University Press.

Chambers, S. A., and T. Carver (2008) *Judith Butler and Political Theory: Troubling Politics*, Oxford: Routledge.

Chan, S. (2011) 'On the uselessness of new wars theory: lessons from African continents', in C. Sylvester (ed.), *Experiencing War*, Abingdon: Routledge, pp. 94–103.

Chang, I. (1997) *The Rape of Nanking: The Forgotten Holocaust of World War II*, New York: Basic Books.

Clausewitz, C. v. (1982 [1832]) *On War*, London: Penguin Classics.

Clifford, L., P. Eichstaedt, K. Glassborow, K. Goetze and C. Ntiryica (2008) 'No sign of end to epidemic', in C. Tosch and Y. Chazan (eds), *Special Report on Sexual Violence in the Democratic Republic of Congo*, The Netherlands: Institute for War and Peace Reporting, October, pp. 4–5.

Cochrane, K. (2008) 'The victims' witness', *Guardian*, 9 May, www.guardian.co.uk/film/2008/may/09/women.congo.

Cohen, D. K. (2010) 'Causes of sexual violence during civil war: cross-national evidence (1980–2009)', Paper presented at the Minnesota International Relations Colloquium.

— (2011) *Explaining Sexual Violence During Civil War*, ProQuest.

Collins, R. (2008) *Violence: A Micro-sociological Theory*, Princeton, NJ, and Oxford: Princeton University Press.

Connell, R. W. (1995) *Masculinities*, Los Angeles: University of California Press.

— (2000) *The Men and the Boys*, Sydney/Cambridge/Berkeley, CA: Allen and Unwin/Polity/University of California Press.

Conrad, J. (1990 [1902]) *The Heart of Darkness*, Dover Thrift Editions.

Cortright, D. (1975) *Soldiers in Revolt. GI resistance during the Vietnam War*, Chicago, IL: Haymarket Books.

Coulter, C., M. Persson and M. Utas (2008) 'Young female fighters in African wars: conflict and its consequences', *Policy Dialogue*, 3, Uppsala: Nordic Africa Institute.

Crewe, E. and E. Harrison (1998) *Whose Development? An Ethnography of Aid*, London and New York: Zed Books.

Crossette, B. (2010) 'A new UN voice calls for criminalizing conflict rape', *The Nation*, 10 September, www.thenation.com/article/154624/new-un-voice-calls-criminalizing-conflict-rape.

Daily Mail (2011) 'Congo named "worst place on earth" to be a woman with more than 400,000 rapes reported in just one year', 12 May, www.dailymail.co.uk/news/article-1386338/Congo-named-worst-place-earth-woman-huge-rate-rapes-revealed.html.

Dauphinée, E. (2007) *The Ethics of Researching War. Looking for Bosnia*, New York: Manchester University Press.

DeGroot, G. J. (2000) 'Introduction to Part 1: Arms and the woman', in G. J. DeGroot and C. Peniston-Bird (eds), *A Soldier and a Woman: Sexual integration in the military*, Harlow: Pearson Education, pp. 3–17.

Delanty, G. (2009) 'The European heritage: history, memory and time', in C. Rumford (ed.), *The Sage Handbook of European Studies*, London: Sage, pp. 36–51.

Derrida, J. (1982) 'Sending: on representation', *Social Research*, 49(2): 294–326.

— (1993) *Aporias: Dying – awaiting
(one another at) the 'limits of truth'*,
Stanford, CA: Stanford University
Press.

Dickinson, E. (2009) 'A bright shining
slogan. How "hearts and minds" came
to be', *Foreign Policy*, 24 August.

Diken, B. and C. Laustsen (2005) 'Becoming abject: rape as a weapon of war',
Body and Society, 11(1): 111–28.

Dillon, M. (2004) 'Correlating sovereign
and biopower', in J. Edkins, V. Pin-Fat
and M. Shapiro (eds), *Sovereign Lives:
Power in Global Politics*, London:
Routledge.

Disch, L. (2003) 'Impartiality, storytelling,
and the seductions of narrative: an
essay at an impasse', *Alternatives:
Global, Local, Political*, 28(2): 253–63.

Dolan, C. (2011) *Social Torture: The Case of
Northern Uganda, 1986–2006*, New York
and Oxford: Berghahn Books.

Doty, R. L. (1993) 'Foreign policy as social
construction: a post-positivist analysis
of US counterinsurgency policy in
the Philippines', *International Studies
Quarterly* 37(3): 367–89.

— (1996) *Imperial Encounters. The Politics
of Representation in North–South
Relations*, Minneapolis: University of
Minnesota Press.

Douma, N. and D. Hilhorst (2012) 'Fond
de commerce? Sexual violence assistance in the Democratic Republic of
Congo', Occasional Paper 2, Wagenin-
gen University.

Duffield, M. (2007) *Development, Security
and Unending War: Governing the World
of Peoples*, Cambridge and Malden,
MA: Polity.

Dunn, K. C. (2003) *Imagining the Congo:
The International Relations of Identity*,
New York: Palgrave Macmillan.

Ebenga, J., and T. N'Landu (2005) 'The
Congolese National Army: in search of
an identity', in M. Rupiya (ed.), *Evolu-
tions and Revolutions: A contemporary
history of militaries in Southern Africa*,

Pretoria: Institute for Security Studies,
pp. 63–84.

EC Investigative Mission (1993) *EC Investigative Mission into the Treatment of
Muslim Women in the Former Yugoslavia:
Report to EC Foreign Ministers*, Copen-
hagen: Ministry of Foreign Affairs.

Edkins, J. (1999) *Poststructuralism and
International Relations: Bringing the
Political Back In*, Boulder, CO: Lynne
Rienner.

— (2003) 'Humanitarianism, humanity,
human', *Journal of Human Rights*,
2(2): 253–8, doi: 10.1080/14754830320
00078224.

— (2005a) 'Ethics and practices of
engagement: intellectuals as experts',
International Relations, 19(1).

— (2005b) 'Exposed singularity', *Journal
for Cultural Research*, 9(4), 359–86, doi:
10.1080/14797580500252548.

Edkins, J., N. Persram and V. Pin-Fat
(1999) *Sovereignty and Subjectivity*,
Boulder, CO: Lynne Rienner.

Ehrenreich, B. (1997) *Blood Rites: Origins
and history of the passions of war*, New
York: Henry Holt.

Elahi, M. (2007) 'War on Congo's women',
Guardian, 25 December.

Elshtain, J. B. (1987) *Women and War*, New
York: Basic Books.

Emmens, S. (producer) (2010) 'Soldiers
armed with lucky charms', sciencemu-
seumdiscovery.com/blogs/collections/
soldiers-armed-with-lucky-charms/.

Enloe, C. (1983) *Does Khaki Become You?
The Militarization of Women's Lives*,
Berkeley: University of California
Press.

— (1990) *Bananas, Beaches and Bases:
Making feminist sense of international
politics*, Berkeley: University of California Press.

— (2000) *Maneuvers: The international
politics of militarizing women's lives*,
Berkeley: University of California
Press.

— (2002) 'Masculinity as a foreign policy
issue', in S. Hawthorne and B. Winter
(eds), *September 11, 2001: Feminist Per-*

spectives, North Melbourne: Spinifex Press, pp. 254–9.

— (2004) 'Demilitarization – or more of the same? Feminist questions to ask in the postwar moment', in C. Enloe (ed.), *The Curious Feminist: Searching for Women in a New Age of Empire*, Berkeley and Los Angeles: University of California Press.

— (2007) *Globalization and Militarism: Feminists make the link*, New York: Rowman and Littlefield.

Eriksson, M. (2010) 'Defining rape. Emerging obligations for states under international law?', PhD dissertation, Örebro University, Kållered.

Eriksson Baaz, M. (2005) *The Paternalism of Partnership: A postcolonial reading of identity in development aid*, London: Zed Books.

— (2011) *The Price for Peace? Military integration and continued conflict in the Democratic Republic of Congo (DRC). Africa Programme Report*, Stockholm: Swedish Defence College.

Eriksson Baaz, M. and M. Stern (2008) 'Making sense of violence: voices of soldiers in the Congo (DRC)', *Journal of Modern African Studies*, 46(01): 57–86.

— (2009) 'Why do soldiers rape? Masculinity, violence, and sexuality in the armed forces in the Congo (DRC)', *International Studies Quarterly*, 53(2): 495–518.

— (2010) *The Complexity of Violence: A critical analysis of sexual violence in the Democratic Republic of the Congo (DRC)*, Stockholm: Sida and Nordic Africa Institute.

— (2011) 'Whores, men and other misfits: undoing "feminization" in the armed forces in the DRC', *African Affairs*, 110(441): 563–85, doi: 10.1093/afraf/adr044.

— (2013a) 'Fearless fighters and submissive wives: negotiating identity among women soldiers in the Congo', Under second round of review at *Armed Forces and Society*.

— (2013b) 'Willing reform? An analysis of defence reform initiatives in the DRC', in A. Bigsten (ed.), *Globalization and Development: Rethinking Interventions and Governance*, New York: Routledge.

Eriksson Baaz, M. and J. Verweijen (2013) 'The volatility of a half-cooked bouillabaisse: rebel–military integration and conflict dynamics in eastern DRC', forthcoming *African Affairs*, June 2013.

Ertürk, Y. (2008) 'Report of Special Rapporteur on violence against women, its causes and consequences', A/HRC/7/6, *Promotion and Protection of All Human Rights, Civil, Political, Economic, Social, and Cultural, Including the Right to Development*, Geneva: Human Rights Council.

Farwell, N. (2004) 'War rape: new conceptualizations and responses', *Affilia*, 19(4): 389–403, doi: 10.1177/0886109904268868.

First Geneva Convention (1949) *First Geneva Convention for the Amelioration of the Condition of the Wounded and Sick in Armed Forces in the Field*.

Foucault, M. (1972) *The Archaeology of Knowledge and the Discourse on Language*, New York: Pantheon.

— (1977) *Discipline and Punish*, London: Penguin.

— (1991) *Discipline and Punishment: The birth of the prison*, trans. A. Sheridan, Harmondsworth: Penguin.

— (2005) *The Order of Things: An Archaeology of the Human Sciences*, London: Routledge.

Fourth Geneva Convention (1949) *Fourth Geneva Convention relative to the Protection of Civilian Persons in Time of War*.

Fromm, E. (1973) *The Anatomy of Human Destructiveness*, New York: Henry Holt.

Fukuyama, F. (2004) *State-building: Governance and World Order in the 21st Century*, New York: Cornell University Press.

Gettleman, J. (2007) 'Rape epidemic raises trauma of Congo war', *New York Times*, 7 October, www.nytimes.com/2007/10/07/world/africa/07congo.html.

— (2009a) 'Symbol of unhealed Congo: male rape victims', *New York Times*, 5 August, www.nytimes.com/2009/08/05/world/africa/05congo.html?sq=rape%20+%20war%20+%20congo&st=nyt&scp=70&pagewanted=print.

— (2009b) 'Clinton presses Congo on minerals', *New York Times*, 11 August.

— (2009c) 'Clinton presents plan to fight sexual violence in Congo', *New York Times*, 12 August, www.nytimes.com/2009/08/12/world/africa/12diplo.html?_r=1.

— (2009d) 'Congo army helps rebels get arms, U.N. finds', *New York Times*, 25 November, www.nytimes.com/2009/11/25/world/africa/25congo.html.

Gikandi, S. (1996) *Maps of Englishness: Writing identity in the culture of colonialism*, New York: Columbia University Press.

Girard, R. (1995) *Violence and the Sacred*, London: Johns Hopkins University Press.

Global Witness (2009) *Faced with a Gun What Can You Do? War and the militarisation of mining in eastern Congo*, London: Global Witness.

Golding, W. (1954) *Lord of the Flies*, London: Faber and Faber.

Goldstein, J. S. (2001) *War and Gender: How gender shapes the war system and vice versa*, Cambridge: Cambridge University Press.

Good Intentions Are Not Enough (n.d.) *Good Donors Are Key to Good Aid*, goodintents.org/.

Gottschall, J. (2004) 'Explaining wartime rape', *Journal of Sex Research*, 41(2): 129–36.

Grisel, M. (2011) *Resolution*, Sweden: Giant Film Production.

Grossman, D. (2009) *On Killing*, revised edn, USA: Back Bay Books.

Groth, A. N. and H. J. Birnbaum (1979) *Men Who Rape: The psychology of the offender*, New York: Plenum Press.

Guardian (2011) 'Congo army colonel guilty of ordering mass rape on New Year's Day', *Guardian*, 21 February.

Guichaoua, Y. (2012a) 'Concluding remarks', in Y. Guichaoua (ed.), *Understanding Collective Political Violence*, Palgrave Macmillan.

— (2012b) 'Introduction: Individual drivers of collective violence and the dynamics of armed groups', in Y. Guichaoua (ed.), *Understanding Collective Political Violence*, Palgrave Macmillan.

Hall, S. (1992) 'The West and the rest. Discourse and power', in S. Hall and B. Gieben (eds), *Formations of Modernity*, Cambridge: Polity in association with the Open University.

— (1996a) 'Introduction: Who needs "identity"?', in S. Hall and P. du Gay (eds), *Questions of Cultural Identity*, London: Sage.

— (1996b) 'When was "the post-colonial"? Thinking at the limit', in I. Chambers and L. Curti (eds), *The Post-Colonial Question. Common Skies, Divided Horizons*, London and New York: Routledge.

— (1997a) 'The spectacle of the "Other"', in S. Hall (ed.), *Representation: Cultural Representations and Signifying Practices*, London and New Delhi: Sage in association with the Open University.

— (1997b) *Representation: Cultural Representations and Signifying Practices*, London and New Delhi: Sage in association with the Open University.

— (1997c) 'The work of representation', in S. Hall (ed.), *Representation: Cultural Representations and Signifying Practices*, London and New Delhi: Sage in association with the Open University, pp. 13–74.

Hansen, L. (2001) 'Gender, nation, rape: Bosnia and the construction of security', *International Feminist Journal of Politics*, 3(1): 55–75.

— (2006) *Security as Practice: Discourse Analysis and the Bosnian War*, New York: Routledge.

Hansson, S. (forthcoming) 'Who brings

the water? Negotiating state responsibility in water sector reform in Niger', PhD thesis, University of Gothenburg, Gothenburg.

Hansson, S., S. Hellberg with M. Stern (eds) (forthcoming) *Studying the Agency of Being Governed*, London: Routledge.

Harvard Humanitarian Initiative and Oxfam (2010) *'Now the world is without me': An investigation of sexual violence in Eastern Democratic Republic of Congo*.

Hatzfeld, J. (2005) *Machete Season: The killers in Rwanda speak*, trans. L. Coverdale, New York: Farrar, Straus and Giroux.

Helliwell, C. (2000) '"It's only a penis": rape, feminism, and difference', *Signs*, 25(3): 789–816.

Henry, N. (2011) *War and Rape: Law, Memory and Justice*, London: Routledge.

Herbert, B. (2006) 'Punished for being female', *New York Times*, 2 November.

— (2009) 'The invisible war', *New York Times*, 21 February.

Herschinger, E. (2010) *Constructing Global Enemies: Hegemony and Identity in International Discourses on Terrorism and Drug Prohibition*, London: Routledge.

Higate, P. (2004) 'Gender and peacekeeping case studies: the DRC and Sierra Leone', *ISS Monograph*, vol. 91, Pretoria: Institute for Security Studies (ISS).

— (2012a) 'Mercenary killer or embodied veteran? The case of Paul Slough and the Nisour Square massacre', in C. Ogden and S. Wakeman (eds), *Corporeality: The Body and Society*, Chester: University of Chester, forthcoming.

— (2012b) 'Foregrounding the in/visibility of military and militarized masculinities', in M. Eriksson Baaz and M. Utas (eds), *Beyond 'Gender and Stir': Reflections on gender and SSR in the aftermath of African conflicts*, NAI Policy Dialogue, Uppsala: Nordic Africa Institute.

Higate, P. and M. Henry (2004) 'Engendering (in)security in peace support operations', *Security Dialogue*, 35(4): 481–98.

Higate, P. and J. Hopton (2005) 'War, militarism and masculinities', in M. S. Kimmel, J. Hearn and R. W. Connell (eds), *Handbook of Studies on Men's Masculinities*, Thousand Oaks, CA: Sage.

Hobbes, T. (1651) *Leviathan*.

Hockey, J. (1986) *Squaddies: Portrait of a Subculture*, Exeter: Exeter University Press.

Hodgson, K. (2003) 'A war on women', *Guardian*, 25 November.

Hoffman, D. (2011) *The War Machines: Young Men and Violence in Sierra Leone and Liberia*, Durham, NC: Duke University Press.

Hoge, W. (2005) 'U.N. relief official condemns use of rape in African wars', *New York Times*, 22 June, query.nytimes.com/gst/fullpage. html?res=9D01E6DC103BF931A15755 C0A9639C8B63.

Horwood, C., J. Ward, C. McEvoy, P. Shipman and L. Rumble (2007) *The Shame of War: Sexual violence against women and girls in conflict*, Nairobi: OCHA and IRIN.

Howarth, D. R., A. J. Norval and Y. Stavrakakis (2000) *Discourse Theory and Political Analysis: Identities, Hegemonies and Social Change*, Manchester and New York: Manchester University Press.

Human Rights Watch (2000) *Federal Republic of Yugoslavia. Kosovo: Rape as a Weapon of 'Ethnic Cleansing'*, New York: Human Rights Watch.

— (2002) *The War within the War: Sexual violence against women and girls in Eastern Congo*, New York: Human Rights Watch.

— (2003) *'We'll Kill You If You Cry.' Sexual Violence in the Sierra Leone Conflict*, New York: Human Rights Watch.

— (2009a) *DR Congo: Brutal Rapes by Rebels and Army*, www.hrw.org/ news/2009/04/08/dr-congo-brutal-rapes-rebels-and-army, accessed 8 May 2012.

— (2009b) *Soldiers Who Rape, Commanders Who Condone*, New York: Human Rights Watch, July.

— (2009c) *You Will Be Punished. Attacks on civilians in Eastern Congo*, New York: Human Rights Watch, December.

Human Rights Watch/Africa and Human Rights Watch Women's Rights Project (1996) *Shattered Lives: Sexual violence during the Rwandan genocide and its aftermath*, London: Human Rights Watch.

Humphreys, F. (2011) 'Sensationalism or silence in the Congo: rape, death and the media', www.worldandmedia.com/drc/sensationalism-or-silence-in-the-congo-rape-death-and-the-media-2605.html.

Hunt, R. N. (2008) 'An acoustic register, tenacious images, and Congolese scenes of rape and repetition', *Cultural Anthropology*, 23(2): 220–53.

Hutchings, K. (2007) 'Feminist ethics and political violence', *International Politics*, 44(1): 90–106.

— (2008a) 'Cognitive short cuts', in J.Parpart and M. Zalewski (eds), *Rethinking the Man Question*, London: Zed Books.

— (2008b) 'Making sense of masculinity and war', *Men and Masculinities*, 10(4): 389–404.

Hypatia (2003) 'Feminist philosophy and the problem of evil', Special issue, *Hypatia*, 18(1).

ICRC (n.d.-a) 'Rule 93. Rape and other forms of sexual violence', *Customary IHL Database*, www.icrc.org/customary-ihl/eng/docs/v1_cha_chapter32_rule93, accessed June 2012.

— (n.d.-b) 'Practice relating to Rule 93. Rape and other forms of sexual violence', *Customary International Humanitarian Law*, www.icrc.org/customary-ihl/eng/docs/v2_rul_rule93, accessed 1 June 2012.

Inayatullah, N. and D. L. Blaney (2004) *International Relations and the Problem of Difference*, New York and London: Routledge.

Institute of International Law (1969) 'The distinction between military objectives and non-military objectives in general and particularly the problems associated with weapons of mass destruction', 9 September 1969, www.icrc.org/ihl.nsf/ FULL/445?OpenDocument, accessed June 2012.

Isango, E. (2003) 'Congolese rebels in cannibal atrocities', *Guardian*, 16 January, www.guardian.co.uk/world/2003/jan/16/congo.unitednations.

Isikozlu, E. and A. S. Millard (2010) *Towards a Typology of Wartime Rape*, Brief 43, Bonn: Bonn International Centre for Conversion.

Jabri, V. (1998) 'Restyling the subject of responsibility in international relations', *Millennium – Journal of International Studies*, 27(3): 591–611.

— (2004) 'Feminist ethics and hegemonic global politics', *Alternatives*, 29(3): 265–84.

— (2007a) 'Solidarity and the spheres of culture: the cosmpolitan and the postcolonial', *Review of International Studies*, 33: 715–28.

— (2007b) *War and the Transformation of Global Politics*, London: Palgrave Macmillan.

Johnson, H. F. (2009) *War's Ultimate Secret Weapon. Stop Rape Now*, 29 September, UN Action Against Sexual Violence in Conflict.

Johnson, K., J. Scott, B. Rughita, M. Kisielewski, J. Asher, R. Ong and L. Lawry (2010) 'Association of sexual violence and human rights with physical and mental health in territories of the eastern Democratic Republic of the Congo', *JAMA*, 304(5): 553–62.

Jones, A. (2000) 'Gendercide and genocide', *Journal of Genocide Research*, 2(2): 185–211.

— (2002) 'Gender and genocide in Rwanda', *Journal of Genocide Research*, 4(1): 65–94.

Kahorha, J. (2011) 'The worst places in the

world for women: Congo', *Guardian*, 14 June.

Kalyvas, S. (2006) *The Logic of Violence in Civil War*, Cambridge: Cambridge University Press.

Kaplan, R. D. (1994) 'The coming anarchy: how scarcity, crime, overpopulation, tribalism, and disease are rapidly destroying the social fabric of our planet', *The Atlantic*.

Kapoor, I. (2004) 'Hyper-self-reflexive development? Spivak on representing the Third World "Other"', *Third World Quarterly*, 25(4): 627–47.

Kassimeris, G. (ed.) (2006) *The Barbarization of Warfare*, London: C. Hurst & Co.

Keen, D. (2005) *Conflict and Collusion in Sierra Leone*, Oxford: James Currey.

Kibasomba, R. (2005) 'Post-war defence integration in the Democratic Republic of the Congo', ISS Paper 119, Pretoria: Institute for Security Studies.

King, A. (2007) 'The word of command: communication and cohesion in the military', *Armed Forces and Society*, 33(4): 493–512.

Kippenberg, J. (2009) *Democratic Republic of Congo: Soldiers Who Rape, Commanders Who Condone: Sexual Violence and Military Reform in the Democratic Republic of Congo*, New York: Human Rights Watch.

Kirby, P. (2012) 'How is rape a weapon of war?: feminist international relations, modes of critical explanation and the study of wartime sexual violence', *European Journal of International Relations*, OnlineFirst, February.

Kirke, C. (2010) 'Orders is orders ... aren't they? Rule bending and rule breaking in the British Army', *Ethnography*, 11(3): 359–80.

Koo, K. L. (2002) 'Confronting a disciplinary blindness: women, war and rape in the international politics of security', *Australian Journal of Political Science*, 37(3): 525–36.

Kort, M. (producer) (2007) *A Conversation with Eve Ensler: Femicide in the Congo*, www.pbs.org/pov/lumo/special_ensler.php.

Kovitz, M. (2003) 'The roots of military masculinity', in P. Higate (ed.), *Military Masculinities. Identity and the State*, Westport, CT, and London: Praeger.

Kristof, N. D. (2008) 'The weapon of rape', *New York Times*, 15 June, www.nytimes.com/2008/06/15/opinion/15kristof.html?sq=rape%20+%20war%20+%20congo&st=nyt&scp=33&pagewanted=print.

— (2010a) 'From "Oprah" to building a sisterhood in Congo', *New York Times*, 3 February, www.nytimes.com/2010/02/04/opinion/04kristof.html.

— (2010b) 'The grotesque vocabulary in Congo', *New York Times*, 10 February, www.nytimes.com/2010/02/11/opinion/11kristof.html.

Lacan, J. (1977) *Ecrits*, London: Tavistock.

— (1993) *The Seminar*, Book III: *The Psychoses 1955–1956*, New York: Norton.

Lacey, M. (2004) 'In Congo war, even peacekeepers add to horror', *New York Times*, 18 December, query.nytimes.com/gst/fullpage.html?res=9D0DE1D81530F93BA25751C1A9629C8B63&pagewanted=all.

Laclau, E. (1990) *New Reflections on the Revolution of Our Time*, London: Verso.

Laclau, E. and C. Mouffe (1985) *Hegemony and Socialist Strategy: Towards a Radical Democratic Politics*, London: Verso.

Landry, D. and G. MacLean (eds) (1996) *The Spivak Reader: Selected Works of Gayatri Chakravorty Spivak*, New York and London: Routledge.

Laudati, A. (2012) *Unearthing the 'Politics of the Belly' from Below: The Role of Peripheral Economies in Eastern DRC*, forthcoming.

Le Carré, J. (2010) 'Hell on Earth: John le Carré on Congo', *Guardian*, 16 January, www.guardian.co.uk/world/2010/jan/16/congo-john-le-carre.

Leatherman, J. L. (2011) *Sexual Violence and Armed Conflict*, Cambridge: Polity.

Left, S. (2005) 'UN reports atrocities in Congo', *Guardian*, 17 March 2005.

Leiby, M. (2009) 'Wartime sexual violence in Guatemala and Peru', *International Studies Quarterly*, 53(2): 445–68.

Lemke, J. L. (1995) *Textual Politics: Discourse and Social Dynamics*, London: Taylor & Francis.

Leonardsson, H. (2012) 'Reports of rape and sexual violence in war. The Guardian and the New York Times 2003–2011', Unpublished manuscript, School of Global Studies, Gothenburg University.

Lewis, D. A. (2009) 'Unrecognized victims: sexual violence against men in conflict settings under international law', *Wisconsin International Law Journal*, 27(1): 1–49.

Li Reviews (2008) 'The Greatest Silence: Rape in the Congo debuts April 8 on HBO', 21 March, www.li-reviews.com/2008/03/21/press-release-greatest-silence/, accessed 5 June 2012.

Littlewood, R. (1997) 'Military rape', *Anthropology Today*, 13(2): 7–16.

Lloyd, M. (2007) *Judith Butler: From norms to politics*, Cambridge: Polity.

Lundahl, M. (2010) 'Kvinnor, Vithet, och de Andras Litteratur', *Tidskrift för genusvetenskap*, 1/2.

Lwambo, D. (2011) *'Before the war, I was a man': Men and masculinities in Eastern DR Congo*, Goma: Heal Africa.

MacCoun, R. J., E. Kier and A. Belkin (2006) 'Does social cohesion determine motivation in combat? An old question with an old answer', *Armed Forces and Society*, 32(4): 646–54.

MacKenzie, M. (2010) 'Ruling exceptions: female soldiers and everyday experiences of civil conflict', in C. Sylvester (ed.), *Experiencing War*, Abingdon: Routledge.

MacKinnon, C. (1989) 'Sexuality, pornography and method: "pleasure under patriarchy"', *Ethics*, 99(2): 314–46.

Masters, C. (2008) 'Bodies of technology and the politics of the flesh', in J. Parpart and M. Zalewski (eds), *Rethinking the Man Question: Sex, gender, and violence in international relations*, London: Zed Books, pp. 87–107.

— (2010) 'Cyborg soldiers and militarized masculinities', *Eurozine*.

— (2012) *Militarism, Gender and (In)Security: Biopolitical Technologies of Security and the War on Terror*, London: Routledge.

McClintock, A. (1995) *Imperial Leather: Race, Gender and Sexuality in the Colonial Conquest*, New York and London: Routledge.

McEwan, C. (2001) 'Postcolonialism, feminism and development: intersections and dilemmas', *Progress in Development Studies*, 1(2): 93–111.

— (2009) *Postcolonialism and Development*, London and New York: Routledge.

McGreal, C. (2006) 'Hundreds of thousands raped in Congo wars', *Guardian*, 14 November, www.guardian.co.uk/world/2006/nov/14/congo.chrismcgreal/print.

— (2007) 'Hundreds of thousands of women raped for being on the wrong side', *Guardian*, 12 November, www.guardian.co.uk/world/2007/nov/12/congo.international.

— (2008) 'Inside the villages where every woman is a victim of hidden war', *Guardian*, 5 December.

Melmot, S. (2008) 'Candide au Congo: L'échec annoncé de la réforme du secteur de sécurité (RSS)', in E. Durand (ed.), *Focus Stratégique*, vol. 9, Paris: Laboratoire de Recherche sur la Défense, L'Institut Français des Relations Internationales (IFRI).

Mendieta, E. (2003) 'Afterword: identities: postcolonial and global', in L. M. Alcoff and E. Mendieta (eds), *Identities*, Oxford: Blackwell, pp. 407–16.

Michalski, M. and J. Gow (2007) *War, Image and Legitimacy: Viewing contemporary conflict*, London and New York: Routledge.

Milbank, D. (2000) 'A candidate's lucky charms. John McCain is hoping superstition will see him through', *Washington Post*, 19 February, www. washingtonpost.com/wp-srv/ WPcap/2000-02/19/067r-021900-idx. html.

Mintzberg, H. (1992) 'Five Ps for strategy', in H. Mintzberg and J. B. Quinn (eds), *The Strategy Process*, Englewood Cliffs, NJ: Prentice-Hall International Editions, pp. 12–19.

Mirage Men (producer) (2010) *RAND, Superstition and Psychological Warfare*, 24 May, miragemen.wordpress. com/2010/11/03/rand-superstition-and-psychological-warfare/.

Mohanty, C. T. (1991) 'Under Western eyes: feminist scholarship and colonial discourses', in C. T. Mohanty, A. Russo and L. Torres (eds), *Third World Women and the Politics of Feminism*, Indianapolis: Indiana University Press.

— (2003) *Feminism without Borders. Decolonizing Theory, Practicing Solidarity*, Durham, NC, and London: Duke University Press.

MONUC Human Rights Division (2007) *The Human Rights Situation in the Democratic Republic of Congo (DRC), during the Period of July to December 2006*, Kinshasa: UN Mission in Democratic Republic of Congo (MONUC).

Moran, M. H. (1995) 'Warriors or soldiers? Masculinity and ritual transvestism in the Liberian civil war', in C. R. Sutton (ed.), *Feminism, Nationalism, and Militarism*, Arlington, VA: American Anthropological Association.

Morgan, D. J. H. (1994) 'Theater of war: combat, the military, and masculinities', in H. Brod and M. Kaufman (eds), *Theorizing Masculinities*, London: Sage.

Moser, C. O. N. and F. C. Clark (eds) (2001) *Victims, Perpetrators or Actors: Gender, armed conflict, and political violence*, London: Zed Books.

Mouffe, C. (2000) 'Politics and passions: the stakes of democracy', *Ethical Perspectives*, 7(2/3): 146–50.

Mudimbe, V. Y. (1994) *The Idea of Africa*, Bloomington: Indiana University Press.

Munn, J. (2008) 'National myths and the creation of heroes', in J. Parpart and M. Zalewski (eds), *Rethinking the Man Question: Sex, gender, and violence in international relations*, London: Zed Books, pp. 143–61.

Muñoz-Rojas, D. and J.-J. Frésard (2004) 'The roots of behaviour in war: understanding and preventing IHL violations', *International Review of the Red Cross*, 853: 189–206.

Muscara, A. (2010) 'Mass gang rape exposes systematic sexual violence', Inter Press Service (IPS), 24 August, ipsnews.net/africa/nota. asp?idnews=52594.

Nagl, J. A. (2005) *Learning to Eat Soup with a Knife: Counterinsurgency Lessons from Malaya and Vietnam*, Chicago, IL, and London: University of Chicago Press.

Neiman, S. (2002) *Evil in Modern Thought: An Alternative History of Philosophy*, Princeton, NJ, and Oxford: Princeton University Press.

Neumann, I. B. (2008) 'Discourse analysis', in A. Klotz and D. Prakash (eds), *Qualitative Methods in International Relations: A Pluralist Guide*, Basingstoke: Palgrave Macmillan, pp. 61–77.

New York Times (1993) 'Rape was a weapon of Serbs, U.N. says', *New York Times*, 19 October, www. nytimes.com/1993/10/20/world/ rape-was-weapon-of-serbs-un-says. html?pagewanted=print.

— (2003) 'U.N. says Congo rebels carried out cannibalism and rapes', *New York Times*, 16 January, www.nytimes. com/2003/01/16/world/un-says-congo-rebels-carried-out-cannibalism-and-rapes.html.

Niarchos, C. N. (1995) 'Women, war, and rape: challenges facing the International Tribunal for the Former Yugoslavia', *Human Rights Quarterly*, 17(4): 649–90.

Nisbet, R. A. (1969) *Social Change and*

History: Aspects of the Western Theory of Development, New York: Oxford University Press.

Nordstrom, C. (2004) *Shadows of War: Violence, Power, and International Profiteering in the Twenty-first Century*, Berkeley, Los Angeles and London: University of California Press.

Nzwili, F. (2009) 'Churches support victims of rape in DR Congo', 27 July, reliefweb. int/node/318362, accessed 5 June 2012.

Ohambe, M. C. O., J. B. B. Muhigwa and B. M. W. Mamba (2005) 'Women's bodies as a battleground: sexual violence against women and girls during the war in the Democratic Republic of Congo, South Kivu (1996–2003)', in M. R. Galloy, N. Sow and C. Hall (eds), London: Réseau des Femmes pour un Développement Associatif (RFDA), Réseau des Femmes pour la Défense de Droits et al Paix (RFDP), International Alert.

O'Malley, P. (1996) 'Risk and responsibility', in A. Barry, T. Osborne and N. Rose (eds), *Foucault and Political Reason. Liberalism, neoliberalism and rationalities of government*, Chicago, IL: University of Chicago Press.

OmniPeace (producer) (2012) *Stamp Out Violence against Women and Girls of the Congo*, 23 May, www.omnipeace.com/ initiatives/stamp-out-violence/.

Onsrud, M., S. Sjøvenian, R. Luhiriri and D. Mukwege (2008) 'Sexual violence-related fistulas in the Democratic Republic of Congo', *International Journal of Gynecology and Obstetrics*, 103(3): 265–69.

Oprah Show (Producer) (2009) 'Lisa Shannon visits the Congo', Video, www. oprah.com/world/Lisa-Shannon-Visits-the-Women-of-the-Congo-Video.

Outhwaite, W. (2009) 'Europe beyond East and West', in C. Rumford (ed.), *The Sage Handbook of European Studies*, London: Sage, pp. 52–68.

Paglia, C. (1992) 'Rape and modern sex war', in *Sex, Art, and American Culture: Essays*, New York: Vintage, pp. 49–54.

Pankhurst, D. (2009) 'Sexual violence in war', in L. Shepherd (ed.), *Gender Matters in Global Politics: A Feminist Introduction to International Relations*, London: Routledge.

Park, J. (2007) 'Sexual violence as a weapon of war in international humanitarian law', *International Public Policy Review*, 3(1).

Parpart, J. (1995a) 'Deconstructing the development "expert": gender, development and the "vulnerable groups"', in M. H. Marchand and J. L. Parpart (eds), *Feminist/Postmodernism/ Development*, London and New York: Routledge.

— (1995b) 'Post-modernism, gender and development', in J. Crush (ed.), *Power of Development*, London and New York: Routledge.

— (2008) 'Masculinity/ies, gender and violence in the struggle for Zimbabwe', in J. Parpart and M. Zalewski (eds.), *Rethinking the Man Question: Sex, gender, and violence in international relations*, London: Zed Books, pp. 181–202.

— (2010) 'Masculinity, gender and the new wars', *Norma*, 2: 86–96.

Pateman, C. and M. L. Shanley (eds) (1991) *Feminist Interpretations and Political Theory*, Oxford: Polity.

Pieterse, J. N. (1992a) *White on Black: Images of Africa and Blacks in Western Popular Culture*, New Haven, CT, and London: Yale University Press.

— (ed.) (1992b) *Emancipations, Modern and Postmodern*, London: Sage.

Pin-Fat, V. (2000) '(Im)possible universalism: reading human rights in world politics', *Review of International Studies*, 26(4): 663–74.

— (forthcoming) *Cosmopolitanism and the End of Humanity: A Grammatical Reading of Posthumanism*.

Pin-Fat, V. and M. Stern (2005) 'The scripting of Private Jessica Lynch: biopolitics, gender, and the "feminization" of the U.S. military', *Alternatives*, 30: 25–53.

Pinauld, C. (2011) 'Understanding war-related violence in Southern Sudan: beyond the gender lines', Paper presented at the ECAS conference, Uppsala.

Pole Institute et al. (2004) 'An open wound: the issue of gender-based violence in North Kivu', *Regards Croisés*, 11.

Polgreen, L. and M. Simons (2007) 'Hague court inquiry focuses on rapes', *New York Times*, 23 May, www.nytimes.com/2007/05/23/world/africa/23car.html?partner=rssnyt&emc=rss.

Polman, L. (2011) *War Games*, London: Penguin.

Pottier, J. (2005) 'Escape from genocide: the politics of identity in Rwanda's massacres', in V. Broch-Due (ed.), *Violence and Belonging: The quest for identity in post-colonial Africa*, Abingdon: Routledge.

— (2007) 'Rights violations, rumour, and rhetoric: making sense of cannibalism in Mambasa, Ituri (DRC)', *Journal of the Royal Anthropological Institute*, 13(4): 825–43.

Pratt, M. and L. Werchick (2004) 'Sexual terrorism: rape as a weapon of war in eastern Democratic Republic of Congo', Assessment Report, Washington, DC: USAID/DCHA.

Price, L. S. (2001) 'Finding the man in the soldier-rapist: some reflections on comprehension and accountability', *Women's Studies International Forum*, 24(2): 211–27.

Requejo, J. (2010) 'Taking stock of maternal, newborn and child survival', Countdown to 2015 Decade Report (2000–2010), World Health Organization and UNICEF.

Richards, P. (1996) *Fighting for the Rain Forest War. Youth and Resources in Sierra Leone*, London: James Currey and Heinemann.

Richey, L. A. and S. Ponte (2011) *Brand Aid. Shopping Well to Save the World*, Minneapolis: University of Minnesota Press.

Riley, D. (1988) '"Am I that name?" Feminism and the category of "women"', in *History*, London: Macmillan Press.

Rotberg, R. I. (2003) *When States Fail: Causes and Consequences*, Princeton, NJ: Princeton University Press.

Royal Air Force Museum (producer) (n.d.) *Aviation Lucky Charms and Mascots*, www.rafmuseum.org.uk/cosford/exhibitions/lucky-charms/.

Said, E. W. (1978) *Orientalism*, New York: Vintage.

Samset, I. (n.d.) Draft manuscript.

Sawyer, I. and A. van Woudenberg (2009) 'You will be punished: attacks on civilians in eastern Congo', in R. Peligal, A. Mawson and J. Ross (eds), *You Will Be Punished: Attacks on civilians in eastern Congo*, New York: Human Rights Watch.

Schott, R. M. (2003) 'Introduction', *Hypatia*, Special issue on 'Feminist philosophy and the problem of evil', 18(1): 1–9, doi: 10.1111/j.1527-2001.2003.tb00776.x.

— (2004) 'The atrocity paradigm', *Hypatia*, 19(4): 204–11, doi: 10.1111/j.1527-2001.2004.tb00157.x.

— (2007) *Feminist Philosophy and the Problem of Evil*, Bloomington: Indiana University Press.

— (2008) '"Just war and the problem of evil"', *Hypatia*, Special issue on 'Just war', 23(2).

— (2011) 'War, rape, natality and genocide', *Journal of Genocide Research*, 13(1/2): 5–21.

Scott, J. (1999) *Gender and the Politics of History*, New York: Columbia University Press.

Second Geneva Convention (1949) *Second Geneva Convention for the Amelioration of the Condition of Wounded, Sick and Shipwrecked Members of Armed Forces at Sea*.

Seifert, R. (1994) 'War and rape: a preliminary analysis', in A. Stiglmayer (ed.), *Mass Rape: The war against women in Bosnia-Herzegovina*, Lincoln: University of Nebraska Press, pp. 54–72.

— (1996) 'The Second Front: the logic of sexual violence in wars', *Women's Studies International Forum*, 19(1): 35–43.

Shannon, L. (2010) 'No, sexual violence is not cultural', *New York Times*, 25 June.

— (2011) *A Thousand Sisters. My Journey into the worst place on earth to be a woman*, Berkeley, CA: Seal Press.

Sharlach, L. (1999) 'Gender and genocide in Rwanda: women as agents and objects of genocide', *Journal of Genocide Research*, 1(3): 387–99.

Shepherd, L. J. (2007) '"Victims, perpetrators and actors" revisited: exploring the potential for a feminist reconceptualisation of (international) security and (gender) violence', *British Journal of Politics and International Relations*, 9(2): 239–56.

Siebold, G. (2007) 'The essence of military group cohesion', *Armed Forces and Society*, 33(2): 286–95.

Simons, M. (2009) 'Long sentences for atrocities in Sierra Leone', *New York Times*, 9 April, www.nytimes.com/2009/04/09/world/africa/09leone.html.

— (2010) 'War crime trials begins for Congolese politician', *New York Times*, 22 November, www.nytimes.com/2010/11/23/world/africa/23hague.html.

Sinha, M. (1995) *Colonial Masculinity*, Manchester: Manchester University Press.

Sivakumaran, S. (2005) 'Male/male rape and the "taint" of homosexuality', *Human Rights Quarterly*, 27(4): 1274–1306.

— (2007) 'Sexual violence against men in armed conflict', *European Journal of International Law*, 18(2): 253–76.

— (2008) 'Male/male rape', Paper presented at the OCHA Seminar Series, New York.

— (2010) 'Lost in translation: UN responses to sexual violence against men and boys in situations of armed conflict', *International Review of the Red Cross*, 92(877).

Sjoberg, L. and C. E. Gentry (2007) *Mothers,*

Monsters, Whores: Women's violence in global politics, London: Zed Books.

Skjelsbaek, I. (2001) 'Sexual violence and war: mapping out a complex relationship', *European Journal of International Relations*, 7(2): 211–38.

— (2010) *The Elephant in the Room. An Overview of How Sexual Violence Came to Be Seen as a Weapon of War*, Oslo: Peace Research Institute Oslo (PRIO).

— (2012) *The Political Psychology of War Rape*, London: Routledge.

Sloane, E. C. (2012) *Modern Military Strategy: An Introduction*, London: Routledge.

Smith, A. D. (2010) 'The doctor who heals victims of Congo's war rapes', *Guardian*, 14 November, www.guardian.co.uk/world/2010/nov/14/doctor-mukwege-congo-war-rapes.

Soeters, J. L., D. J. Winslow and A. Weibull (2006) 'Military culture', in G. Caforio (ed.), *Handbook of the Sociology of the Military*, New York: Springer, pp. 237–54.

Sontag, S. (2003) *Regarding the Pain of Others*, New York: Farrar, Straus and Giroux.

Spivak, G. C. (1988) 'Can the subaltern speak?', in C. Nelson and L. Grossberg (eds), *Marxism and the Interpretation of Culture*, Chicago: University of Illinois Press, pp. 271–315.

— (1990) *The Post-Colonial Critic: Interviews, Strategies, Dialogues*, New York: Routledge.

— (1993) *Outside in the Teaching Machine*, London: Routledge.

— (1999) *A Critique of Postcolonial Reason: Toward a Critique of a Vanishing Present*, Cambridge, MA: Harvard University Press.

— (2004) 'Righting wrongs', *South Atlantic Quaterly*, 103(2/3).

Stanley, P. (1999) 'Reporting of mass rape in the Balkans: plus ça change, plus c'est la même chose? From Bosnia to Kosovo', *Civil Wars*, 2(2): 74–110, doi: 10.1080/13698249908402408.

Statement by the President of the Security Council (2004) S/PRST/2004/46, 14 December.

— (2005) S/PRST/2005/25, 21 June.

Staub, E. (1989) *The Roots of Evil: The origins of genocide and other group violence*, Cambridge: Cambridge University Press.

— (2011) *Overcoming Evil: Genocide, violent conflict and terrorism*, New York: Oxford University Press.

Stearns, J. (2009a) 'Is celebrity activism useful?', 10 November, congosiasa. blogspot.com/2009/11/is-celebrity-activism-useful.html.

— (2009b) 'Are we focusing too much on sexual violence in the DRC?', 14 December, congosiasa.blogspot. com/2009/12/are-we-focusing-too-much-on-sexual.html.

— (2011) 'A bar fight leads to pillage and rape in Fizi', 9 January, congosiasa. blogspot.com/2011/01/bar-fight-leads-to-pillage-and-rape-in.html.

Stern, M. (2005) *Naming Security – Constructing Identity: 'Mayan women' in Guatemala on the eve of 'peace'*, Manchester: Manchester University Press.

— (2006) '"We" the subject: the power and failure of (in)security', *Security Dialogue*, 37(2): 187–205.

— (2011) 'Gender and race in the European security strategy: Europe as a "force for good"?', *Journal of International Relations and Development*, 14: 28–59.

Stern, M. and M. Nystrand (2006) *Gender and Armed Conflict*, Sida.

Stern, M. and M. Zalewski (2009) 'Feminist fatigue(s): reflections on feminism and familiar fables of militarisation', *Review of International Studies*, 35(3): 611–30.

Stiglmayer, A. (ed.) (1994) *Mass Rape: The war against women in Bosnia-Herzegovina*, Lincoln: University of Nebraska Press.

Storr, W. (2011) 'The rape of men', *Guardian*, 17 July, www.guardian.co.uk/society/2011/jul/17/the-rape-of-men/print.

Strachan, H. (2007) *Clausewitz in the Twenty-first Century*, New York: Oxford University Press.

Sudetic, C. (2011) 'Congo justice: unintended consequences', blog.soros. org/2011/04/congo-justice-unintended-consequences/.

Swiss, S. and J. E. Giller (1993) 'Rape as a crime of war. A medical perspective', *Journal of the American Medical Association*, 270(5): 612–15, doi: 10.1001/jama.1993.03510050078031.

Syed, J. and F. Ali (2011) 'The white woman's burden: from colonial civilization to Third World development', *Third World Quarterly*, 32(2): 349–65.

Sylvester, C. (1994) 'Empathetic cooperation: a feminist method for IR', *Millennium – Journal of International Studies*, 23: 315–34, doi: 10.1177/03058298940230021301.

— (1999) 'Development studies and post-colonial studies: disparate tales of the "Third World"', *Third World Quarterly*, 20(4): 703–21.

— (2007) 'Whither the international at the end of IR?', *Millennium – Journal of International Studies*, 35(3): 551–71.

— (ed.) (2011) *Experiencing War*, Abingdon: Routledge.

Third Geneva Convention (1949) *Third Geneva Convention relative to the Treatment of Prisoners of War*.

Thornhill, R. and C. T. Palmer (2000) *A Natural History of Rape: Biological bases of sexual coercion*, Cambridge, MA: MIT Press.

Titeca, K. (2011) 'Access to resources and predictability in armed rebellion: the FAPC's short-lived "Monaco" in eastern Congo', *Africa Spectrum*, 46(2): 43–70.

Torfing, J. (2004) 'Discourse theory: achievements, arguments and challenges', in D. R. Howarth and J. Torfing (eds), *Discourse Theory in European Politics: Identity, Policy and Governance*, Basingstoke: Palgrave.

Trefon, T. (2011) *Congo Masquerade. The Political Culture of Aid Inefficiency and*

Reform Failure, London and New York: Zed Books.

Trinh, T. M.-h. (1989) *Women, Native, Other: Writing Postcoloniality and Feminism*, Bloomington: Indiana University Press.

Turner, G. (2004) *Understanding Celebrity*, London: Sage.

UN Action (2007) *UN Action Against Sexual Violence in Conflict: Stop Rape Now*.

— (2010) *UN Action Against Sexual Violence in Conflict. Information Brochure: Stop Rape Now*.

— (2011) *UN Action Against Sexual Violence in Conflict. Progress Report 2010–2011: Stop Rape Now*.

UN Commission on Human Rights (1996) *C.H.R. res. 1996/71: Situation of human rights in the Republic of Bosnia and Herzegovina, the State of Bosnia and Herzegovina, the Republic of Croatia and the Federal Republic of Yugoslavia (Serbia and Montenegro)*.

— (2005) *Human Rights Resolution 2005/85: Technical Cooperation and Advisory Services in the Democratic Republic of the Congo*, E/CN.4/RES/2005/85.

UN General Assembly (1995) *A/RES/49/196. Situation of human rights in the Republic of Bosnia and Herzegovina, the Republic of Croatia and the Federal Republic of Yugoslavia (Serbia and Montenegro)*, A/RES/49/196 C.F.R.

— (1996) *A/RES/50/192. Rape and abuse of women in the areas of armed conflict in the former Yugoslavia*, A/RES/50/192 C.F.R.

— (1997) *A/RES/51/115. Rape and abuse of women in the areas of armed conflict in the former Yugoslavia*, A/RES/51/115. C.F.R.

— (2006) *A/RES/60/170. Situation of human rights in the Democratic Republic of the Congo*, A/RES/60/170 C.F.R.

UN Group of Experts (2004–11) *Reports of the Group of Experts submitted through the Security Council Committee established pursuant to resolution 1533 (2004) concerning the Democratic Republic of the Congo*.

UN OCHA (2008) *The Nature, Scope and Motivation for Sexual Violence against Men and Boys in Armed Conflict*, UN OCHA research meeting, 26 June.

UN Security Council (2009) *Final Report of the Group of Experts on the Democratic Republic of the Congo*, vol. S/2009/603, New York: United Nations.

UNDP (2008) 'Sexual violence against women and children in armed conflict', Parliamentary hearing at the United Nations.

UNFPA and B. Crossette (2010) 'State of world population 2010. From conflict and crisis to renewal: generations of change', in UNFPA (ed.), *The United Nations Population Fund*.

UNIFEM, Stop Rape Now and United Nations (2010) *An Analytical Inventory of Peacekeeping Practice*.

United Nations (1980) *Convention on Prohibitions or Restrictions on the Use of Certain Conventional Weapons which May be Deemed to be Excessively Injurious or to Have Indiscriminate Effects*, adopted 10 October 1980, treaties.un.org/pages/ViewDetails. aspx?src=TREATY&mtdsg_no=XXVI-2&chapter=26&lang=en, June 2012.

— (1994) *Final Report of the Commission of experts. Established Pursuant to Security Council Resolution 780*, UN document S/1994/674.

UNSC (2000) *Women and peace and security*, S/RES/1325.

— (2008) *Women and peace and security*, S/RES/1820.

— (2009a) *Women and peace and security*, S/RES/1888.

— (2009b) *Children and armed conflict*, S/RES/1882.

Utas, M. (2005) 'Victimcy, girlfriending, soldiering: tactic agency in a young woman's social navigation of the Liberian war zone', *Anthropological Quarterly*, 78(2): 403–30.

— (2008) 'Abject heroes: marginalised youth, modernity and violent pathways of the Liberian civil war', in J. Hart (ed.), *In Years of Conflict: Adoles-*

cence, *political violence and displacement*, Oxford: Refugee Studies Centre and Berghahn Books, pp. 111–38.

— (2011) 'Victimcy as social navigation: from the toolbox of Liberian child soldiers', in A. Özerdem and S. Podder (eds), *Child Soldiers: From recruitment to reintegration*, London: Palgrave Macmillan, pp. 213–30.

Utas, M. and M. Jörgel (2008) 'West Side Boys: military navigation in the Sierra Leone civil war', *Journal of Modern African Studies*, 46(3): 487–511.

V-Day (n.d.) 'Eve Ensler', www.vday.org/about/more-about/eveensler, accessed 30 May 2012.

Van Creveld, M. (2008) *The Changing Face of War: Combat from the Marne to Iraq*, New York: Ballantine.

Vaughan-Williams, N. (2007) 'Beyond a cosmopolitan ideal: the politics of singularity', *International Politics*, 44(1): 107–24.

Verweijen, J. (2013a) *The ambiguity of militarization. The complex interaction between the Congolese armed forces and civilians in the Kivu provinces, eastern DRC*, Doctoral dissertation in Conflict Studies, Utrecht University, forthcoming.

— (2013b) 'Half-brewed: the lukewarm results of creating an integrated Congolese military', in R. Licklider (ed.), *New Armies from Old. Merging Competing Military Forces after Civil War*, forthcoming.

Viner, K. (2011) 'City of joy: new hope for Congo's brutalized women', *Guardian*, 9 April, www.guardian.co.uk/world/2011/apr/09/city-of-joy-congo-women-rape.

Vlassenroot, K. and T. Raeymaekers (2009) 'Kivu's intractable security conundrum', *African Affairs*, 108(432): 475–84.

Wallström, M. (2010a) '"Conflict minerals" finance gang rape in Africa', *Guardian*, 14 August, www.guardian.co.uk/commentisfree/2010/aug/14/conflict-minerals-finance-gang-rape.

— (2010b) *Statement to the Security Council of the Special Representative of the Secretary-General on Sexual Violence in Conflict*.

— (2011) 'Ending sexual violence: from recognition to action', *New Routes*, 16(2): 49–52.

Weil, S. (1965) 'The Iliad, or the poem of force', *Chicago Review*, 18(2).

Weiner, A. (2006) 'Something to die for, a lot to kill for: the Soviet system and the barbarisation of warfare 1939–1945', in G. Kassimeris (ed.), *The Barbarization of Warfare*, London: C. Hurst & Co.

Weinstein, J. M. (2007) *Inside Rebellion: The politics of insurgent violence*, Cambridge: Cambridge University Press.

Weldes, J., M. Laffey, H. Gusterson and R. Duvall (eds) (1999) *Cultures of Insecurity: States, communities, and the production of danger*, Minneapolis: University of Minnesota Press.

Weston, K. (2002) *Gender in Real Time: Power and Transience in a Visual Age*, New York: Routledge.

Whitehead, S. M. (2002) *Men and Masculinities: Key themes and new directions*, Cambridge: Polity.

Whitworth, S. (2004) *Men, Militarism, and UN Peacekeeping: A gendered analysis*, Boulder, CO: Lynne Rienner.

Wiegman, R. (2003) *Women's Studies on Its Own: A next wave reader in institutional change*, Durham, NC: Duke University Press.

— (2004) 'On being in time with feminism', *Modern Language Quarterly: A Journal of Literary History*, 65(1).

Wink, W. (1992) *Engaging the Powers: Discernment and Resistance in a World of Domination*, Minneapolis: Fortress Press.

Wodak, R. (2008) 'Introduction: Discourse studies – important concepts and terms', in R. Wodak and M. Krzyzanowski (eds), *Qualitative Discourse Analysis in the Social Sciences*, Basingstoke: Palgrave Macmillan.

Woman Stats Project (n.d.) *Woman*

Stats Project, www.womanstats.org/, accessed 13 June 2012.

Wong, L. (2006) 'Combat motivation in today's soldiers: US Army War College Strategic Studies Institute', *Armed Forces and Society*, 32(4): 659–63.

Wood, E. J. (2009) 'Armed groups and sexual violence: when is wartime rape rare?', *Politics and Society*, 37(1): 131–62.

— (2010) 'Sexual violence during war: variation and accountability', in A. Smeulers (ed.), *Collective Violence and International Criminal Justice*, Antwerp: Intersentia.

— (2012) 'Rape during war is not inevitable: variation in wartime sexual violence', in M. Bergsmo, A. B. Skre and E. J. Wood (eds), *Understanding and Proving International Sex Crimes*, Beijing: Torkel Opsahl Academic EPublisher, pp. 389–419.

Woodward, R. and T. Winter (2007) *Sexing the Soldier: The politics of gender and the contemporary British Army*, New York: Routledge.

York, G. (2010) 'Anti-rape fund in Congo wasted: critics', *Globe and Mail*, www. theglobeandmail.com/news/world/anti-rape-funds-in-congo-wasted-critics/article1500360/.

Young, C. and T. Turner (1985) *The Rise and Decline of the Zairian State*, Madison: University of Wisconsin Press.

Young, R. J. C. (1995) *Colonial Desire: Hybridity in Theory, Culture and Race*, London and New York: Routledge.

Yuval-Davis, N. (1997) *Gender and Nation*, London: Sage.

Zalewski, M. (2012) *Feminist International Relations: 'Exquisite Corpse'*, London: Routledge.

Zarkov, D. (1997) 'War rapes in Bosnia: on masculinity, femininity and power of the rape victim identity', *Tijdschrift voor Criminologie*, 39(2): 140–51.

Zehfuss, M. (2007) 'Subjectivity and vulnerability: on the war with Iraq', *International Politics*, 44(1): 58–71, doi: 10.1057/palgrave.ip.8800158.

Zimbardo, P. G. (2008) *The Lucifer Effect: Understanding How Good People Turn Evil*, New York: Random House.

Žižek, S. (2009) *Violence: Six Sideways Reflections*, London: Profile Books.

Index

153

About Zed Books

Zed Books is a critical and dynamic publisher, committed to increasing awareness of important international issues and to promoting diversity, alternative voices and progressive social change. We publish on politics, development, gender, the environment and economics for a global audience of students, academics, activists and general readers. Run as a co-operative, Zed Books aims to operate in an ethical and environmentally sustainable way.

Find out more at:

www.zedbooks.co.uk

For up-to-date news, articles, reviews and events information visit:

http://zed-books.blogspot.com

To subscribe to the monthly Zed Books e-newsletter, send an email headed 'subscribe' to:

marketing@zedbooks.net

We can also be found on **Facebook**, **ZNet**, **Twitter** and **Library Thing**.

www.ingramcontent.com/pod-product-compliance
Lightning Source LLC
Chambersburg PA
CBHW031137270326
41929CB00011B/1658